P9-ARW-129

THE BIG BANG OF NUMBERS

ALSO BY MANIL SURI

The Death of Vishnu

The Age of Shiva

The City of Devi

THE BIG BANG OF NUMBERS

How to Build the Universe Using Only Math

MANIL SURI

W. W. NORTON & COMPANY
Independent Publishers Since 1923

Printed in the United States of America
First Edition

For information about special discounts for bulk purchases, please contact W. W. Norton Special Sales at specialsales@wwnorton.com or 800-233-4830

Manufacturing by Lakeside Book Company
Book design by Lovedog Studio
Production manager: Julia Druskin

ISBN: 978-1-324-00703-6

W. W. Norton & Company, Inc., 500 Fifth Avenue, New York, N.Y. 10110
www.wwnorton.com

W. W. Norton & Company Ltd., 15 Carlisle Street, London W1D 3BS

1 2 3 4 5 6 7 8 9 0

In grateful memory of Professor M. S. Huzurbazar, who told me to listen because mathematics was calling my name.

CONTENTS

Day 3: ALGEBRA

Day 4: PATTERNS

Day 5: PHYSICS

Day 6: INFINITY

Day 7: EMERGENCE

THE BIG BANG OF NUMBERS

Introduction

THE POPE MADE ME WRITE THIS BOOK

On Monday, September 13, 2013, the *New York Times* published my opinion piece "How to Fall in Love with Math." I awoke to find my email inbox overflowing with messages, not just from acquaintances but also from a bewildering number of strangers. Many more responses poured in online on the *Times* website—some enthusiastic about math, some scathing, all impassioned—clearly, I'd struck a nerve. By mid-afternoon, the number of posts had reached 360, and the paper closed the comments section. The article quickly climbed to the top of the *Times*'s most-emailed list and remained there for much of the next day.

The aim of my piece was to challenge the popular notion that mathematics is synonymous with *calculation*. Starting with arithmetic and proceeding through algebra and beyond, the message drummed into our heads as students is that we do math to "get the right answer." The drill of multiplication tables, the drudgery of long division, the quadratic formula and its memorization—these are the dreary memories many of us carry around from school as a result.

But what if we liberated ourselves from the stress of finding "the right answer"? What would math look like if delinked from this calculation-driven motivation? What, if anything, would remain of the subject?

The answer is *ideas*. That's what mathematics is truly about, the realm where it really comes alive. Ideas that engage and intrigue us as humans, that help us understand the universe. Ideas about the perfection of numbers, the nature of space and geometry, the spontaneous formation of patterns, the origins of randomness and infinity. The neat thing is that such ideas can be enjoyed without needing any special mathematical knowledge or being a computation whiz.

This is what I'd observed over the past decade and a half, during which, in addition to my day job as a mathematics professor, I'd been pursuing a dual career as a novelist. The juxtaposition put me in frequent contact with artists, writers, composers, journalists, and I was struck by the curiosity they expressed about math. Some had been good at it, but lost contact with the subject once they chose their career path; others had encountered difficulties learning it and viewed it as an unfulfilled intellectual challenge. Often, I was asked to give a talk not on my writing but about mathematics. "Something really exotic," a few would add, their eyes shiny with daring, as if venturing into an Indian restaurant and asking for the menu's hottest curry.

So I began talking about the mysteries of infinity (a topic that's spicy, but not overwhelmingly so), which eventually developed into an animated PowerPoint talk. I'd go to dinner at people's houses and once the plates had been cleared, ask if they'd like to see the show. (You know you've become a math evangelist when you carry such presentations on a flash drive in your wallet.) These activities got headier, more addictive. I started seeing myself as Florence Nightingale, administering math to the mathless; Johnny Appleseed, scattering math seeds like fairy dust everywhere I went. Some of my targets may have regarded me more like the Ancient Mariner and themselves as the cornered wedding guest. A few had to be rescued by their parents.

My novels (on India, not math) were doing well, so I was able to infiltrate even more venues where nonmathematicians congregated. My coolest coup was at the 2006 Berlin International Literature Festival, where a class of eleventh graders who thought they'd hear me speak about my second novel got my infinity talk instead. They seemed to like it—or at least sat through attentively, without fidgeting (the fact they were German may have had something to do with this).

By 2013, I'd begun to get a sense of the limits of such outreach efforts. So when my *New York Times* op-ed took off, I wondered if I'd finally hit it big. Nothing I'd ever written had ever gone "viral"—in fact, I wasn't even sure what numbers earned that characterization. By Thursday, my piece had climbed into the weekly top-ten list; by Friday, it had inched up a few slots more. Over the weekend, I watched obsessively as it crept into the top three, then nudged its way into second place. What I barely noticed was that the pope had chosen that very week to make some startlingly progressive statements about gays, abortion, and birth control. Just as I was about to claim my rightful pinnacle of victory, he appeared behind me from nowhere, bounding up the list in twos and threes. Quads flexing, cassock billowing, he made one final spectacular jump, to leapfrog clear over me and land in my number one spot.

Now, you may wonder if I developed a lingering grudge against the pope, if I've written this book to vindicate myself in an imagined *mano a mano* with him. Let me assure you that's not the case. I've completely forgiven him and will even be mailing him an autographed copy of the finished book at the Vatican to show no hard feelings remain.

However, his surprise appearance did have a crucial effect: it focused my attention on religion. In the popularity contest with mathematics, religion had handily won—as it almost surely would each time. What did it offer that math didn't? What lesson could one take away for math to draw people in, to compete in the attention economy we live in?

There's no shortage of answers to this question, but I was reminded of a quote I'd seen years earlier that had cut me to the quick. It was attributed to Rob Fixmer, a former editor of the *New York Times*, who was attempting to explain why math got so little media attention:

> Mathematics has no emotional impact. What physicists do challenges people's notions of origins and creations. Math doesn't challenge any fundamental beliefs or what it means to be human.

My immediate reaction was indignation—how could anyone malign something I loved this way? In time, I realized this might be an opinion many shared. Also, though the quote had compared math with physics, the same could be said while comparing math to religion. After all, both physics and religion seek to address the Big Questions—albeit from opposing perspectives: Where does everything come from? Why is the universe the way it is? How do we fit in? The two camps have been duking it out over the answers for centuries, begetting even more attention for themselves.

Math doesn't seem to have a dog to enter in this fight. The subject is abstract, agnostic—ready to describe and analyze phenomena, without having a position of its own to stake. That's the image perceived by most, anyway. Without such blockbuster spectacles as Genesis or the big bang, no wonder math has difficulty competing in the engagement sweepstakes.

But I'm here to tell you this picture of math is inaccurate. Math *does* have a compelling "origins" story, one that creates its basic building blocks out of nothing. With a little inventiveness, this narrative can be extended to show how with these building blocks—called *numbers*—the entire universe could plausibly be constructed. Big Questions do indeed get addressed along the way, with answers that come not from God or science but mathematics.

As I watched my article begin its descent on the *NYT* list, I realized that, as follow-up, I should write precisely the above kind of narrative. One that would flesh out my article's central assertion about math being more about ideas than calculation (which meant I'd need to severely limit formulas and equations!). Something that would not only convey the aesthetic pleasure of the subject but also reveal the deeper connections we—and our cosmos—have with it.

✳

The thought experiment

The book you (and, God willing, the pope) are about to read emerged from that day's realization. Its premise is this. I'm going to put you, the reader, in the driver's seat and have you take on the task of creating the universe using only numbers, and the mathematics you formulate from them. We'll launch this adventure in the next chapter with the above-mentioned "origins" blast—math's very own creation spectacle!

Sitting at the controls of this thought experiment, you'll soon find yourself devising arithmetic, then geometry, then algebra, then physics—all in response to the needs of your universe-in-progress. (This will incidentally answer the question "Why does algebra exist in the world?" asked by untold legions of unhappy schoolchildren.) The perspective you'll get is unusual, even radical: math as the life force of the universe, a top-down driving power that fashions everything that exists. This turns on its head the traditional way mathematics is understood. Rather than regarding it as something we devise to explain preexisting real-life phenomena (given to us by God or physics), we will view mathematics as the fundamental source of creation, with reality trying to follow its dictates as best it can.

Such a view is actually not new—it has precedents traceable all the way back to the ancient Greeks, particularly Plato. What differentiates us from Plato is that we don't assume all of mathematics already exists in some idealized form somewhere, waiting to be discovered, as he did. Rather, we *invent* math—from scratch, and through active, energetic exploration. Math that will *create* the universe, rather than *explain* something already in place.

There are several advantages to this reverse, hands-on approach. For one, it will enable you to get a firsthand taste of the playful nature of mathematics. This is something mathematicians often rhapsodize about but outsiders can find hard to access. Sitting in the driver's seat, you'll see how even simple arithmetic operations like addition and multiplication are, at heart, games. You'll be able to experiment with such games and ideas creatively, as if playing with an abstract set of toy building blocks or Lego bricks. Each time the inevitable question—"What good is this, anyway?"—comes up, the answer will be right there. After all, the components of the universe will, quite literally, be arising from your play!

Contrast this with the alternative, of starting with real-life phenomena and demonstrating how math can be used to approximately model them. Such efforts (as I've noticed in my own outreach) can come across as a "good for you" vitamin, with playfulness often smothered under the weight of technical elaboration. Using play and exploration to bring out the usefulness of math, as we do here, makes the connection feel more natural, effortless. The fact that we can embed math in a single continuous narrative helps show how its different areas are linked.

Another advantage of our approach is that it allows us a fresh look at the "unreasonable effectiveness of mathematics" in describing the universe (as Nobel laureate Eugene Wigner put it). This is a riddle that's central to the subject—how can something so abstract be so uncannily adept at explaining the reality we live in? Clearly, if we demonstrate that the mathematics in our thought experiment leads inevitably to the creation of *everything* in our universe (and only *our* universe, rather than some different one!) then we've come a long way toward changing "unreasonable" to "*very* reasonable" effectiveness.

Let me be up-front: our thought experiment won't quite achieve this kind of slam dunk. Trying to build everything just with math is, to put it mildly, a tad ambitious. However, proceeding step by step, we'll find out what other ingredients might be minimally needed, while getting to appreciate just how deeply numbers are hardwired into our experience. In particular, our flipped perspective will lead us to an insightful new interpretation of how Nature fits in. We'll view her as a building contractor, who has the task of turning our design for the universe into tangible reality. By casting her in this role, we'll be able to better understand why she doesn't follow our mathematical instructions to the letter, what role randomness plays.

Our explorations will also reveal that the universe we live in isn't the only one possible. That's because such basics commonly taken for granted—like size, distance, space—arise innately from mathematics. Consequently, you can make them strikingly different in any universe you create by defining them in alternative mathematical ways. For instance, we'll use the very physical art of crocheting to figure out the different fabrics possible for the universe's geometry. (It turns out that we don't quite know which is the correct one even for our own universe!) Expect to have your "fundamental beliefs" challenged as we explore such variations, thereby checking off one of the boxes in Fixmer's quote.

Another critical question raised in the quote was whether math has "emotional impact." It's true that mathematics, much more than art or music, is experienced more intellectually than viscerally. However, comprehension is often followed by a eureka moment, which is part of the emotional punch math packs. That's what our thought experiment is set up to deliver—whether through games that suddenly open up to reveal a deeper truth, or through "eye candy" fractals that transform into essential drivers of the universe when you engage with the math behind them. Look for such experiences particularly in the section on patterns, where we attempt to make the Mona Lisa prettier through a mathematical makeover.

We've adopted our top-down version of mathematics because it works so well in our exposition. But could this reflect reality? Could math truly be what guides our universe? We debate both sides of this ahead, but the section on physics contains evidence supporting our viewpoint. You'll see how mathematics actually *determines* physical principles like Newton's law of gravitation (and even a very basic idea used in general relativity), rather than just being a language used to explain them.

The penultimate section, on infinity, transports our thought experiment to a more expansive, philosophical realm. While previous sections bring out how mathematics informs such essential qualities as randomness, symmetry, and beauty, we now explore what it tells us about even deeper questions. Does knowledge have boundaries? Is omniscience possible? Can one list successive instants of time? The concept of infinity is essential for delving into such issues. Even though it's something that can only be envisaged, never physically attained, infinity is inextricably tied up with setting down the universe's blueprints; much of the reality around us involves the push and pull of the finite versus the infinite. Be on the lookout for a fictional interplanetary battle through which we'll steer our thought experiment to unlock infinity's secrets.

In the final section, we get down to the task of physically building our universe from the designs we've created. Iteration, where one state in time evolves to another, is the key mathematical process that facilitates construction. We explore how elementary rules of iteration can lead to complex, self-arising phenomena—fractal patterns, stripes on a zebra, "herd intelligence" in ants. Such *emergence* (as the spontaneous generation of complexity is called) could plausibly create life itself.

The issue of creation, of course, is also in the domain of religious philosophy, whose relationship to science we comment on at several points in the text. One of the most essential questions both ask is this: Why do we exist? Is it a result of randomness or intent?

This is the final enigma we address. Mathematics will reveal the answer at the end of our thought experiment.

Ground rules

The most stringent test of our ideas would be if we scrubbed our minds clean of all knowledge of both math and our universe, and started from a blank slate. We're going to do nothing of the sort. Which means that, consciously or unconsciously, such knowledge might seep in to inform everything we create. Will this damage the integrity of our experiment?

It won't. Many mathematical concepts can indeed be deduced

quite naturally from the universe's components (e.g., a sphere from the universe's round objects, the Golden Ratio from sunflower seeds and pine cones). Although we'll often point out such links to familiarize you with a new concept, our charge will be to also find an *independent* path to develop the concept, one that ignores connections with the physical universe, and uses only the math we've created. In other words, our first guideline will be to always strive to maintain our top-down philosophy, by composing a narrative in which math drives the universe, rather than the universe providing us the math.

As far as prior knowledge of math goes, our second rule will be that we can't use a mathematical concept before we've developed it. For instance, we can't arrange numbers along a straight line if we haven't created lines yet. This will ensure we're actually constructing all the math we need as we go along.

Since we're following the math, we can expect (as mentioned earlier) for it to lead us to several alternative universes. We'll look into some of these variations, since they can provide interesting insights into our own reality. However, when a choice arises on how to proceed with our thought experiment, our guideline will be to pick the path most likely to lead us back to our own universe, since that's the one we want to build.

Finally, let's note that thought experiments have a particularly rich tradition in physics and philosophy, where they typically explore rather focused questions or concepts ("What does quantum superposition have to say about the health of Schrödinger's cat?"). We can afford to be more relaxed, since our thought experiment is just the vehicle—the true purpose of our journey is the understanding and appreciation of math. Consequently, we'll take some liberties—step outside our experiment at will, get Nature and the numbers to roam around as characters, have the pope materialize from time to time to offer commentary, and let a Greek chorus of physicists do the same.

✳

How to read this book

As you've no doubt surmised, I'm aiming this book not only at novices but also at math enthusiasts. This includes inquiring high schoolers, teachers and parents, seasoned STEM readers, all the way up to hardened math professionals. If you've ever tried to get anyone excited about math, I promise you'll find new strategies within these pages.

However, a word of warning—some people's experiences with mathematics may have convinced them they can never find its ideas engaging. ("So—good for you, you're moved by the pure beauty of math. Please, take a cookie and congratulate yourself" is how one reader of my *NYT* article put it. Others were less generous.) We can all be resistant to the charms of different endeavors; my biggest blind spot—a problem, since I grew up in India—happens to be cricket (unlike everyone around me, I was always loath to watch it). So, try keeping an open mind as you embark on this book, with the perspective that you *can* enjoy math under the right circumstances. (I'll meanwhile think about giving another shot to cricket.)

Next, for novices, relax. Some of you might find it liberating to read about mathematical ideas for the first time without the nagging worry: Is this going to be on the test? Experience this math appreciation at your own pace—you can always come back to topics that you decide to just skim. Remember, even though the formulas I've used are both elementary and minimal in number, the ideas expressed can run rather deep.

Do read the footnotes, since they're interesting (that's why I added them!). Endnotes, which contain references, technical qualifications, and proofs, but also some engaging elaborations, can sometimes be more mathematically challenging to read. I'll alert you only to the most essential ones as we go along. That said, if you're a mathematician and feel something's missing, chances are it's addressed in one of the unflagged endnotes, so do check them.

You'll notice I've organized this book in "days," as in Genesis. It's a metaphor I couldn't resist, once I realized our mathematical creation would have exactly seven stages. Although these stages don't correspond to what God creates in the Bible (light on Day 1, sky on Day 2, etc.), both progressions suggest how naturally each creation might lead to the next—in our case, arithmetic giving rise to geometry, geometry to algebra, and so on. Math becomes more of a story, not just like Genesis, but also like the narrative of how everything gets built up from the big bang in physics. And who doesn't love a story, as they say.

This brings me to my final note. Both religion and science might have, for each reader, their own place. Despite any apparent irreverence, my aim isn't to supplant or disrespect either. Rather, it's to take you on a thought experiment that introduces another worldview, one fashioned around mathematics. It's an offering that even the pope, my most treasured potential reader, shouldn't have a problem with.*

---

* This is a good spot to make clear that Pope Francis (who's been the pope while I've been writing this book) has not participated in this work in any way beyond his chance entanglement with me on the *NYT* "most emailed" list. The comments, actions, and thoughts I attribute to him ahead are fictional, though details about his educational background are correct.

Day 1

ARITHMETIC

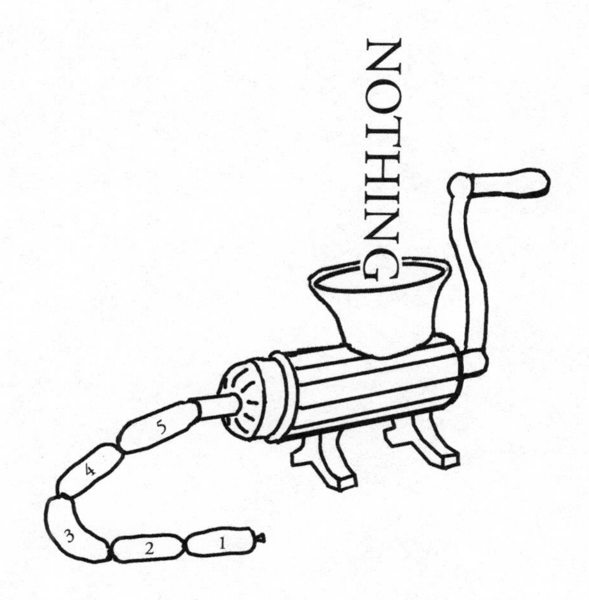

Creating numbers out of nothing

1.

SETTING UP THE BIG BANG

There's a hiccup with the very first step of creating our building blocks, the numbers. Why waste time on "creating" something so obvious? Can't we just assume the numbers exist, and take it from there? After all, even the God of Genesis comes equipped with numbers—witness the way the six days of creation are counted off, with a rest on Day 7. And surely physics also assumes numbers' existence—otherwise one couldn't have the precise values of physical constants needed for the universe to work. Are mathematicians so arrogant that they're going to nitpick with both science and God? Surely this is an exercise as unnecessary as—to put it in my terms—watching cricket?

Fine. I was hoping to give you some insight into how mathematicians operate, but we can skip this step. Provided you just answer one question. What is a number?

That's easy. "One, two, three . . . ," you start, but these are just examples. "A value used to count" or "A symbol used to quantify" are only descriptions—not particularly satisfying ones at that, since they use other undefined terms like "count" and "quantify." Notice how hard it is to actually pin down the definition of a number. It's my second-favorite question to quell an uppity math class. (My favorite is "Define math.")

The problem is we conflate the identity of numbers too much with their function, associating them only with counting and measurement. Perhaps this is a legacy from our early ancestors, who, after all, were motivated by purely practical needs when they embarked on their arduous quest for numbers. Asking them to characterize their invention might have elicited different replies through the ages: notches on a wolf bone, knots on a cord, hieroglyphics and pictographs, all the way to the Indo-Arabic symbols used now. They'd have dismissed us as precious (or worse, demented) had we persisted in asking them to define the essence of a number.

Plato would be an exception. He believed numbers to be abstract entities, living in their own idealized universe—a universe containing galaxies of perfect concepts that exist independently of the material realm. The most humans could hope for was to get occasional glimpses of this cosmos, coming away with discoveries they could use in their own world.

Like Plato, we need to understand the nitty-gritty of numbers. We need to understand the material we're building our universe from.* Unlike him, we won't take the easy way out by assuming numbers are givens, in existence since time immemorial. Rather, we'll *construct* them ourselves, much like an artisanal builder who personally bakes every brick for a house. That's the scout's motto for mathematicians: be prepared to build everything from scratch. As promised, there *will* be a blast!

So imagine you're at a starting point that predates the universe—*all* universes, even Plato's. God has not yet commenced building and neither has physics—in fact, both are waiting somewhat testily for

---

* Plato's building blocks were not numbers but geometric solids, identified with the basic elements of the universe (the tetrahedron was fire; the cube was earth; the octahedron, air; the icosahedron, water; the dodecahedron, the universe as a whole). He expounded on this in *Timaeus*, where he also aimed to explain the formation of the universe, as we do here. However, he did this based on philosophical and theological principles, rather than mathematical ones, such as we're using. Interestingly, Plato's choice of the cube as the earth element has been shown to have some truth to it. The most natural way rocks fracture when broken up is indeed, on average, into six-sided cube-like chunks.

you to give them the numbers first. Given their penchant for sudden eruptions, you'd better get to work.

Except you literally have *nothing* to work with. No energy, no matter, no symbols, no objects to count. You can't even imagine yourself suspended alone in empty space, because space hasn't been invented yet. Even the *concept* of space, along with that of length or area or dimension, has yet to be formulated. Time is hazy as well—perhaps it's waiting to be initiated. You've never felt so isolated, never seen a "nothing" so absolute, so stark. A "nothing" you realize is your only resource—the commodity you'll have to construct your numbers from.

Now, King Lear may have warned that nothing can beget only nothing, but both religion and physics have been enchanted by the opposite, utopian, notion. *Creatio ex nihilo* is the ultimate magic trick, whereby oceans, skies, planets, the universe, life itself, are all created out of nothing (*nihilo*). Either with the help of a divine entity, or unaided, through a spacetime "singularity" (as physicists might put it). Every religion has its own variant—if the universe isn't being blown out by Brahma in a single breath, then it's rising from the infinite sea of chaos posited by the ancient Egyptians.

The reason I call *creatio ex nihilo* a magic trick is that the "nothing" you start with always needs an asterisk (a singularity for the big bang, a Supreme Creator for Genesis). Some of these asterisks can be quite pronounced—making the *nihilo* part essentially a con. What's remarkable about the version you're about to use for the numbers is that the asterisk needed is so trifling, so close to nothing itself. It's called the *empty set*. Let's start by understanding what this is.

The empty set

By *set* we'll simply mean a collection of objects. For instance, we could pictorially denote the set of letters in the word "set" as shown. The nature of the boundary or arrangement of the objects inside doesn't matter, just the fact that they belong.

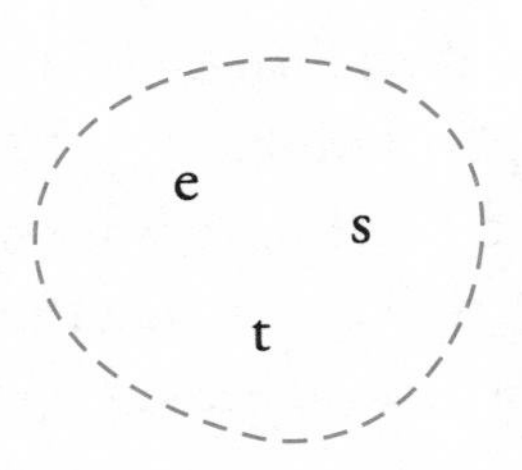

You can think of all sorts of sets: the set of digits greater than 2 but less than 8, the set of cats in your house, the set of trees in your yard.

What happens, though, if you don't have any cats or trees? That's when you get the *empty* set—conveniently visualized as a boundary with nothing enclosed.

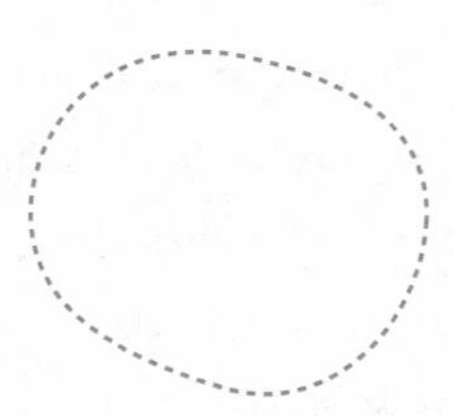

As other examples, consider the set of negative numbers greater than 2, the set of six-headed cats, or (gratuitous swipe) the set of truthful politicians.

Now here's where *creatio ex nihilo* kicks in: numbers can be generated from the empty set. To do so in a stand-alone, mathematically useful way, we need to transcend our cultural training. Think of numbers as sets—in fact, begin by defining the "number" zero as the empty set.

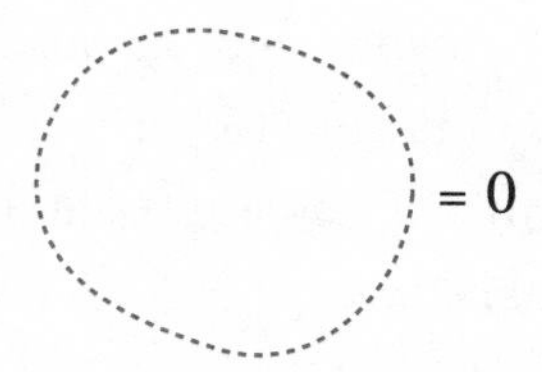

This might feel strange, but it fills a gap—it pins down zero precisely, invests it with an independent mathematical identity. In fact, it's essentially how the Hindus invented zero—they gazed into the void, a mainstay of Hindu and Buddhist philosophy, and realized it could be represented as a number.

Zero happened to be the final digit the Hindus needed, since they already had the others from their counting experience. For us, the process is reversed—we need to start at zero and build from the bottom up. So how do we create the number *one*?

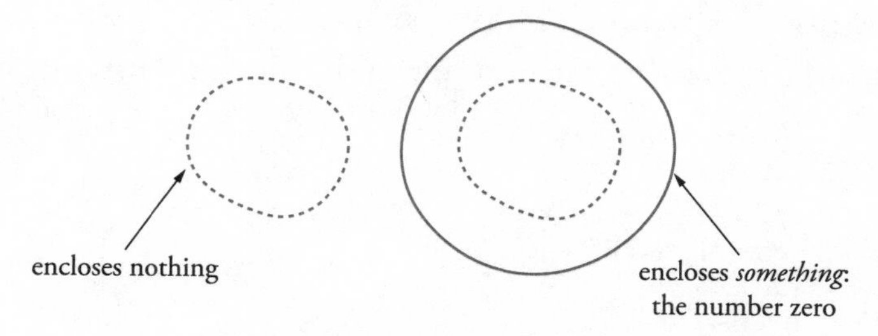

To do this, imagine a new solid boundary around the dotted boundary that represents the empty set. As before, the dotted boundary encloses nothing. But notice that the solid boundary *does* enclose something: the "number" zero we've just created. This solid boundary therefore represents a new set—one that is no longer empty. Define the number *one* to be this new set, that is, one is the set which contains the previous number zero.

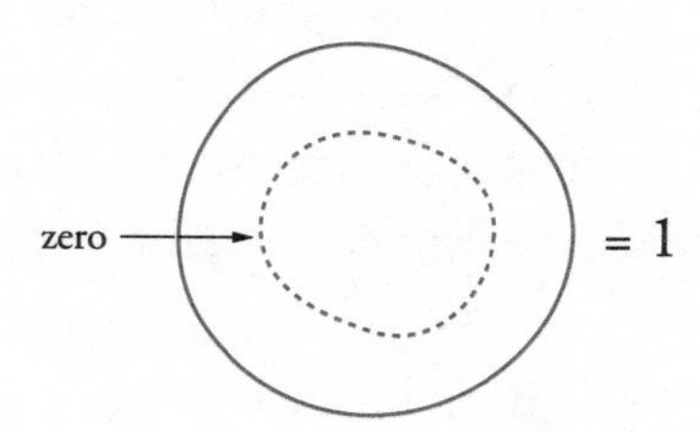

Continue this way. Having figured out how to get from zero to one, we use the same procedure to get from one to its successor. All we have to do is visualize a new set, containing both zero and one, and call this new number *two*.

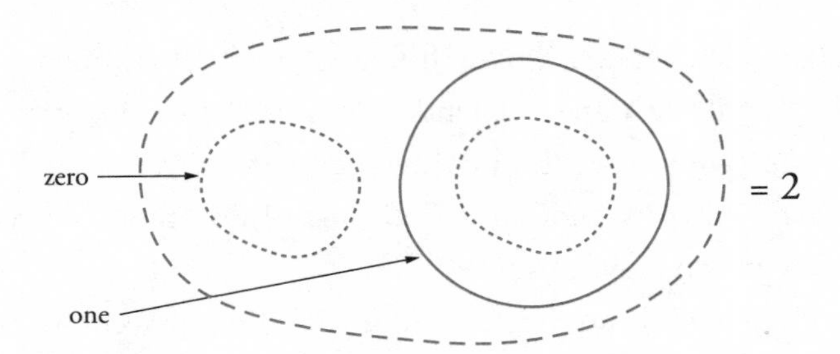

From two to *three*, three to *four*—we can simply make new sets each time, containing all the numbers already created. It's like igniting the fuse to a chain reaction—one that gives successors over and over again (try creating the next few in the process illustrated below).

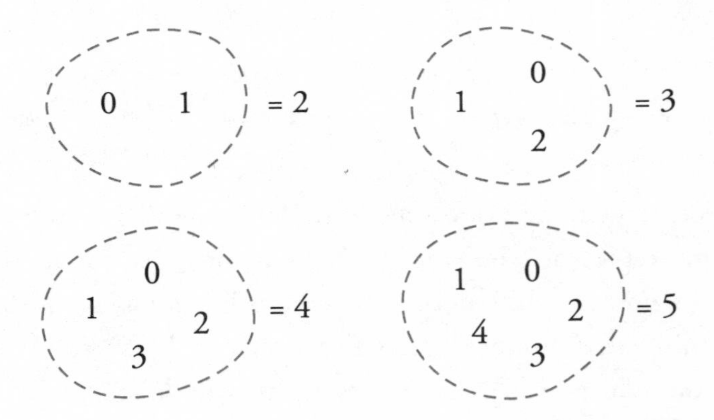

You might object that these creations have little resemblance to the numbers we use in real life for counting. However, notice that "two" is a set containing exactly *two* elements—the numbers zero and one. "Three" is a set containing exactly *three* elements—zero, one, two. A little like a die's face value being indicated by the number of dots on it.

What's the point of all this? For one, it gives you a glimpse into how mathematicians think, building up the subject brick by brick with such elemental definitions (just as physicists might with subatomic particles). More importantly, these "defining sets" give you something to latch on to when thinking of numbers. They're a demonstration that numbers exist as their own independent entities, just as Plato asserted.

So let's birth them, make them the first inhabitants of the universe you create. Let's have that kickoff explosion I proposed (remember the chain-reaction fuse that was lit?). Imagine yourself adrift in nothingness, corralling it somehow into an empty set, astounded at the newborn Zero you've just created. Tinkering around some more, delivering One and Two and Three, still uncertain about this process you've stumbled upon. The Big Bang of Numbers catches you by surprise—a spontaneous blast, like something a youngster experimenting with a chemistry set might ignite. The numbers burst forth, each one generating the next. Minuscule stars, fiery from your act of creation—they streak through your lonely universe, filling it with light and companionship.

2.

GAMES NUMBERS PLAY

YOU'RE DOING RATHER WELL FOR SOMEONE WHO'S just delivered a litter that's countless. Even God, if given credit for every being that's ever existed, may have had only a finite number of offspring. To top it off, your feat of conception was clearly immaculate. As creator, you should be pretty pleased with yourself.

But there's a problem. Your numbers are too abstract, and born into a bit of a nebulous jumble. Wouldn't it be nice if they had personalities to make them more real—even to set them apart, like ordinary children? Hold this thought as we hear from a student of mine who might be of help.

Numbers are people too, my friend

Lili has something unusual to share in a math seminar I'm teaching. Numbers always appear pigmented in their own unique colors for her: one is orange, two purple, three pink, and so on. The technical term for her condition is synesthesia, she informs us, and it can take many forms. Instead of seeing colors when viewing letters or numbers, some might associate fragrances with sounds, others might even be able to "taste" words. For her, the association is so strong that coloring books traumatized her as a child—each time a region labeled one wasn't orange or two wasn't purple, she'd cringe at the unnaturalness of it all. I feel instantly guilty about my color choices chalking numbers in class. Apologies, Lili, for making you squirm—much as teachers enjoy occasionally torturing their students, this wasn't on purpose.

Lili tells us of an extra dimension to her synesthesia: numbers also come with their own genders and personalities. For instance, One is a hard-working single mom, Two is her shy son who might be gay, Three is a sweet and nurturing elder sister, while Seven is a cool dude—the kind with a backward baseball cap, sunglasses, and a skateboard. It can get as hectic as a television sitcom, Lili says.*

Clearly, numbers are alive for Lili in a way they're not for most of us. Could you take a cue from her, to better see numbers as entities in themselves? Maybe even ascribe personas, consistent with the roles they'll play in the mathematical universe to come?

Start with Zero, who we'll make, quite arbitrarily, a "she." (See the endnote on genders and pronouns I've used.) Not only is Zero the originator of an entire tribe, but she's a thinker, a *yogini*, forever pondering the deepest mysteries of numberdom. How does she differ from *nothing*, for instance? What exactly is this entity called a set? Are you truly her creator, and if so, what dispensation allowed you

---

* Interestingly, she's not alone in experiencing numbers with personalities—there's even been a play called *Numesthesia* written about her kind of synesthesia.

to harness emptiness? *Creatio ex nihilo*, as I cautioned, always comes with an asterisk, and your firstborn is relentless in trying to zero in (so to speak) on the trick involved. Her motives, though, are pristine—she's driven by introspection, not mischief.*

In contrast with Zero's meditative nature, One—or Uno, as he prefers to be called—is practical and down-to-earth, a real doer. Were numbers human, he'd be the kid you'd shoot hoops or (horrors) play cricket with. He takes charge of the rest of the brood, tinkers with both his own and others' defining sets. Ever on the lookout for problems to solve using calculation, One—sorry, Uno—is always eager to forge ahead with math.

Two has little of his older brother's brashness. He's placid, stable, good at forging compromises, a cherisher of noble qualities like symmetry and evenness. He does chafe a tiny bit, however, at always being in the shadow of Uno's accomplishments.

Such anthropomorphizing is fun, and we'll keep it handy to help us understand some key ideas ahead. Viewing numbers as denizens of your universe will breathe new life into addition, multiplication, and other such operations. That said, you might already see that assigning *every* number its own personality isn't going to work. The tribe is too vast—even Lili's synesthesia stops at ten.

Fortunately, we don't need personalities to set these denizens apart, since each already comes with a prominent identifying feature—its number (of course!). In fact, this feature is important in terms of not only mathematics but also language. Let's take a short detour to see how it gives rise to some basic semantics.

---

* Zero's questions are all valid issues that come up for anyone laying the logical foundations of mathematics. In most cases, one has to address them through technical assumptions or "axioms"—for example, the axiom "The empty set exists." We'll take a closer look at axioms when we explore geometry on Day 2. For now, let us mention that unexpected problems can arise if one defines sets "naively" without axioms, as we have done (see endnote).

Numbers and comparison

"More," "less," "big," "small"—such terms have no meaning in your universe yet, since size and counting haven't been invented. Numbers so far are just defining sets, floating around like bits of mathematical protoplasm. It's only after you attach appropriate context to them that you'll be able to use them to make comparisons, and ascribe meaning to such terms.

How to attach such context? Recall the numbers emerged in a definite *order* from their Big Bang. So, call a number "less than" or "greater than" another depending on who came first in this progression. Three is less than Five, whom it preceded, but greater than One, whom it succeeded; Six is greater than Four but less than Nine, and so on. Once you get adept at this, you can even start developing a subjective sense of calling a number "big" or "small," based on how many other numbers lie between it and Zero. (Note, though, that numbers you call "big" might become "small" when you compare them with even bigger numbers, so these adjectives are relative.)

Long and short, near and far, rich and poor—numbers will help you define all these concepts, as you proceed with your universe. All derive from the sequence in which numbers succeeded each other during birth.

The ability to compare, however, is a double-edged sword. Once you possess it, the idea of inequality is born. This is a notion that will mesmerize humans (once they appear) in droves. The desire to best will become one of the driving obsessions of civilization, unleashing monstrous inequities in your world. Inequities that could arguably be traced back to and blamed on numbers.

So much so that perhaps the story of Adam and Eve should be rewritten. Perhaps it was the forbidden tree of numeracy, not knowledge, that got them banished from Eden. Perhaps the serpent's offering was a zero, not an apple.

Addition and multiplication

But we're getting way ahead in your creation process. So let's return to the *naturals*—the name* you give the numbers 0, 1, 2, . . . Now that you've produced them, what next? You're probably eager to deploy them in some spectacular feat of creation that will astound both physicists and the pope.

Unfortunately, that's not how math's going to work. At least not in this early stage, when your numbers can't be applied to anything—for the simple reason that nothing else exists! Instead, let's take a page from the typical view mathematicians have of their subject: that it's really a form of play. One invents a bunch of rules as one might for tic-tac-toe or sudoku or Scrabble, and sees where they lead, what deductions can be made from them. The English mathematician G. H. Hardy said it most famously: not only did he call mathematics a game but he also declared being enamored most of all with the game's uselessness!

Except it doesn't necessarily stay that way. Over time, even the most whimsical game might develop practical uses. Hardy himself has been spinning around at quite a clip in his grave for decades, since the parts of his own research he treasured most for their uselessness turned out to be the most flagrantly applicable (e.g., in cryptography).

You're going to have to hope for something similar. For now, give up the vision of erecting the universe's grand monuments. Instead, invent some games to play with the naturals and cross your fingers something comes of it.

This is where giving the numbers life pays its dividend. Imagine them dart and race around as they might for Lili, my synesthetic student. In your unformed universe, though, there's literally nowhere to

---

* The name "naturals" can be traced back to the French mathematician Nicolas Chuquet, who referred to the "natural progression" of numbers 1, 2, 3, . . . in a 1484 treatise. Some omit 0 from the list, but as we've seen, everything follows from 0, so it's *natural* to include it. (Those who omit 0 from the naturals sometimes characterize the numbers 0, 1, 2, . . . as constituting a new set called the "wholes.")

go, nothing to see. The numbers quickly get bored with such limited opportunities for play. Some of the more rambunctious ones start roughhousing among themselves.

You're forced to intervene to restore order. As you're about to pull apart Three and Five, you pause. Why do they remind you of Eight, grappling with each other like that? You realize it's because if you count up Three's defining set, and keep going with Five's, you end up with exactly the same number of elements as in Eight's defining set. Congratulations! You've just discovered the "game" of *addition*.

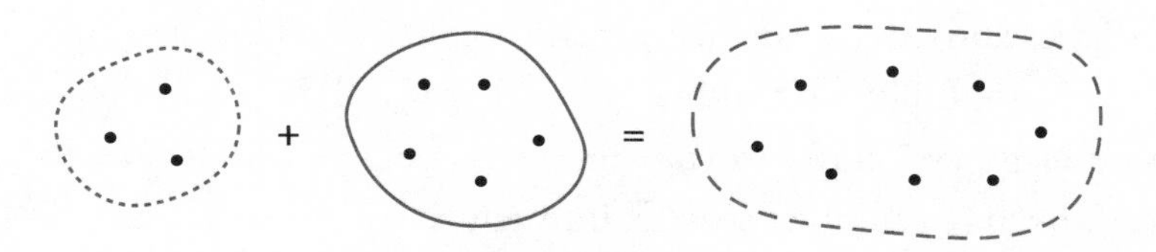

You start entertaining yourself (and your numbers!) with this game. Input two natural numbers (Zero, One, Two, etc.) as players and you get another natural number as the outcome. You discover some interesting facts. First, whether you take the "sum" Three + Five or Five + Three, the answer, Eight, remains the same. (Picture what the above sketch would look like for Five + Three instead of Three + Five.) More strikingly, you can play with *any* pair of numbers—the game always culminates successfully, in the sense that there's always a number waiting in your set of naturals to assume the role of the sum.

You soon cook up another game. This one's played by adding together copies of the *same* number. You know One plus One gives Two, Four + Four + Four gives Twelve, Five + Five + Five + Five gives Twenty. It becomes tedious to list long sums involving the same number, so why not abbreviate such strings? Six + Six + Six . . . , repeated ten times, gives Sixty, which you shorten to "Six times Ten equals Sixty." And then you simply start writing this as

$$\text{Six} \times \text{Ten} = \text{Sixty}.$$

You call this *multiplication*—a game that again depends on pairs of numbers as input.

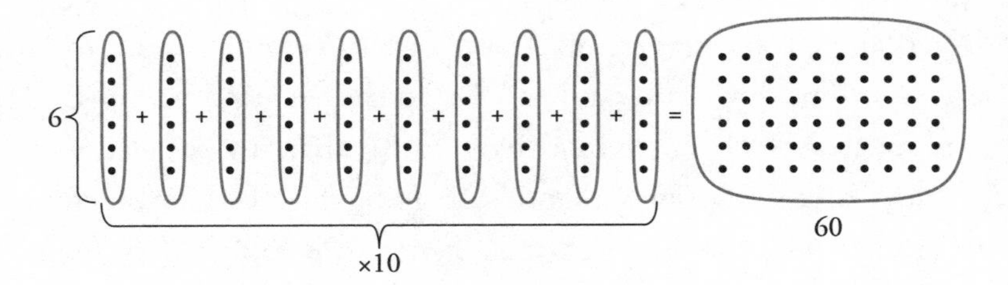

In time, you discover tricks for calculating "products," and invent multiplication tables. You're especially entertained by watching small numbers multiply each other to vault toward bigger targets. The games bring you closer to your numbers, help you understand them better. Even if you don't see any immediate use for such play.

Representing numbers

It turns out the multiplication game *does* end up having a practical use: it yields a convenient way to represent numbers. So far, you've just been calling them by their names, "Zero," "Five," "Twenty," and so on. You've also come up with symbols to represent the first few of them—the usual digits 0 to 9, say. What you'd like is to design a separate such symbol for each of your building blocks, perhaps □ for Ten, Δ for Eleven, ☺ for Twelve, and so on. But this seems impossible, since the numbers are endless. Not only would you run out of ideas, but nobody could remember such a limitless lineup of symbols.

Multiplication comes to the rescue. You notice that Twenty-one equals Two times Ten plus One, while Thirty-six equals Three times Ten plus Six. Writing these observations out, you see a natural method of representation emerge:

$$\text{Twenty-one} = 2 \times \text{Ten} + 1 = 21,$$

Thirty-six = 3 × Ten + 6 = 36.

In other words, the multiplication by Ten and the subsequent addition of units can be treated as implicit steps left unstated. With this convention, you need only the ten symbols for Zero through Nine to be able to represent *any* number. For instance, Fifty-four, which is 5 × Ten + 4, can be represented as 54, and Seventy-three, which is 7 × Ten + 3, as 73. Interpret your own age this way, and it reveals how many complete decades—plus extra years over—you've lived.

If you suspect the choice of using multiples of Ten as your base is arbitrary, you're correct. For a while, you try experimenting with using other numbers as base as well.* But for some reason, Ten seems just right to use as base—not too big and not too small.† Of course, humans started using this base because they had ten fingers. But in this thought experiment, where humans don't exist as yet, we'll refrain from justifying our choice by appealing to this fact.‡

* As an example, with base Eight, you can write Sixteen = 2 × Eight + 0 = 20 and Twenty = 2 × Eight + 4 = 24. If this looks confusing, it's because we're so used to base Ten. Fine to not dwell on this; we won't be using it.

† The smallest natural number you can use as base is Two, which leads to numbers being represented entirely as strings of zeros and ones. This "binary" representation is the one of choice in many computer systems.

‡ The Greeks, under Pythagoras, had a different reason for considering Ten special. They noted that ten points could be arranged in four rows to form an equilateral triangle, which they considered metaphysically auspicious.

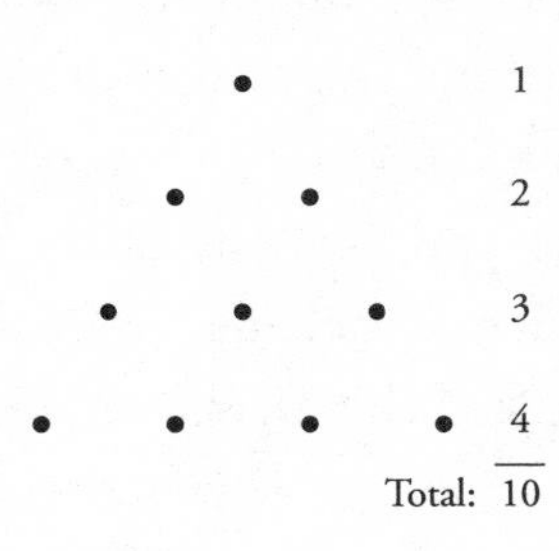

What about the symbol for Ten itself? Using the convention you've developed, you write

$$\text{Ten} = 1 \times \text{Ten} + 0,$$

so Ten = 10. Also,

$$\text{Hundred} = 10 \times \text{Ten} + 0,$$

so you represent it as 100.

The perfection of the naturals

Let's take a moment to savor the aesthetics of what you've created. This, after all, is one of our goals—to get a feel for the so-called elegance of math. Hard as it might be to believe, there's beauty to be gleaned even from something you've plugged through thousands of times before—operations as humdrum as addition, as mundane as multiplication.

Consider how your naturals are both complete and self-contained. No matter how you add or multiply them, the outcome is always another natural just like them. Players are never in danger of crossing a boundary and stepping off the universe by virtue of their sum or product getting too big (something that would happen had there been a largest natural). There are also no unfilled holes that could pose a hazard during play (as might have happened had a number like 10 or 20 been somehow skipped in the creation process). Every possible number that could ever be an outcome is included.

I want to impress on you how special such a situation is in our world. For instance, chemical elements are the building blocks of matter, but not every pair of chemicals can be made to combine (e.g., sodium reacts with chlorine, but it won't with potassium). Or consider Scrabble. Try placing a random string of letters on the board, and you're likely to encounter a "hole"—a word not in the dictionary, and hence inadmissible.

Your naturals, on the other hand, appear perfect. They seem to have no omissions or limits. The games you've invented proceed without a hitch on their seamless playing field. The system runs as smoothly as well-oiled machinery. It therefore comes as a shock when you see this perfection blown away.

3.

MORE GAMES

THE CHAPTER TITLE MIGHT BE A BIT FRUSTRATING. Games are fine and everything, but don't you have a universe to build? Physicists reading this book will probably smirk at how little you've managed to accomplish without their intervention. The pope's going to ring for his secretary and ask just why this book was put on his reading list.

It's true that "games" might suggest frivolity, a sense of frittering time away. But few more effective paths to learning or exploration exist. Ask any teacher who's tried motivating a class, and they'll echo this. What better way to get really good at addition or multiplication, for instance, than to make a game of it?

We've already seen how games helped us discover some of the properties of the naturals. Let's use them now to make more explorations. Our goal is to perfect your building blocks before employing them in the universe's construction. Playing with games is the best way to test the numbers, to uncover gaps or holes where something else might be needed.

The perils of subtraction

The inadequacy of the naturals is revealed by your third game, which you devise as an antidote to addition. Say you've started with 2 and added 3 to get 5. You now want to reverse the process and get back to 2. So in your new game, 3 disengages from 5 to yield 2, which you denote as

$$5 - 3 = 2.$$

You call this game *subtraction.*

The numbers, ever restless for something new, are quick to seize upon this latest diversion. They begin zipping back and forth between addition and subtraction:

$$2 + 3 = 5,$$

followed by

$$5 - 3 = 2,$$

repeated like a two-step dance, over and over again. But then you notice something unexpected: 2 – 5 has no answer waiting in the set of naturals, and neither does 3 – 5. As a result, the defining sets of such pairs get horribly entangled when they try to play this game, with components unable to disengage. You start seeing failed subtractions floating around everywhere, pairs of numbers locked in intractable embrace. As more numbers are sucked into this game, the universe comes to a standstill.

How to unlock your universe? It dawns on you that you may have to create more numbers. That way, for example, the unanswered subtraction 0 − 1 could equal one of these new numbers—you'd call it Minus One. But every number requires a defining set,* and this proves to be a hurdle. You putter around with the defining sets for Zero and One, but there are no obvious modifications that would yield a set for Minus One.

Meanwhile, you realize this same Minus One might also work as the answer to another stuck subtraction: 1 − 2. In fact, you could use it to liberate all such locked pairs involving two successive naturals: 2 − 3, 3 − 4, and so on. You start gathering together such pairs, a painstaking task. There's a mess of unconsummated subtractions from which you have to tweeze out just the ones you want. Once you finish, you herd them all into a set you label −1.

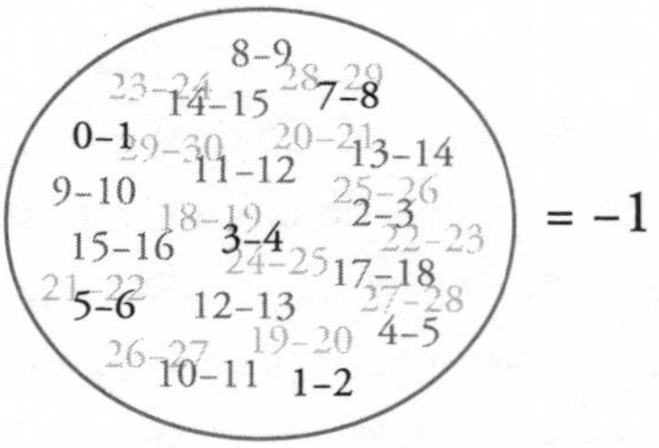

8−9
23−24 14−15 28−29 7−8
0−1 29−30 20−21 13−14
9−10 11−12 25−26
18−19 2−3
15−16 3−4 22−23
24−25 17−18
21−22 5−6 12−13 27−28
19−20 4−5
26−27 10−11 1−2
= −1

You're contemplating your handiwork when you have a sudden thought. Could this be the defining set you were after? The more you think of it, the more you're convinced it's true. It may look nothing like previous specimens, but you've just formulated the defining set for Minus One. It consists of every subtraction where the second natural number is one more than the first.

This recipe is easy to extend. You corral together all the differences 0 − 2, 1 − 3, 2 − 4, . . . , where the second number is *two* more than

---

* Since we're baking each brick from scratch, we want to nail down each new number's construction with the same care with which we defined Zero. Defining sets can get rather abstract, but don't worry. You just have to endure them for a bit, since we'll say goodbye to them at the end of this chapter.

the first. This set, you name the negative number –2. Proceeding this way, you similarly create –3, –4, –5, and so on.

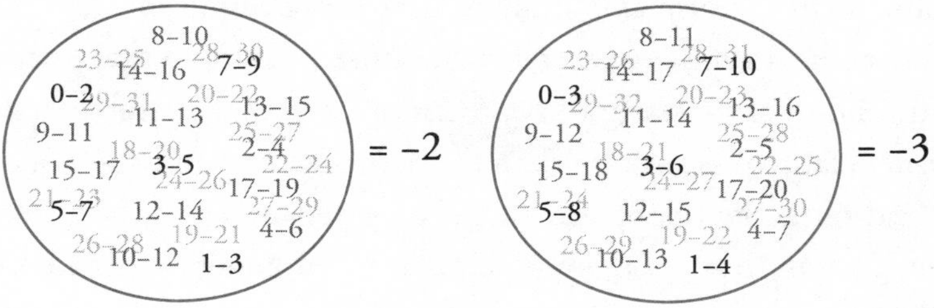

8–10 23–25 28–30 14–16 7–9 0–2 29–31 20–22 13–15 11–13 9–11 25–27 18–20 2–4 15–17 3–5 22–24 24–26 17–19 21–23 5–7 12–14 27–29 26–28 19–21 4–6 10–12 1–3 = –2

8–11 23–26 28–31 14–17 7–10 0–3 29–32 20–23 13–16 11–14 9–12 25–28 18–21 2–5 15–18 3–6 22–25 24–27 17–20 21–24 5–8 12–15 27–30 26–29 19–22 4–7 10–13 1–4 = –3

With these new numbers at your disposal, you can end the subtraction gridlock. You go around matching frozen pairs to the appropriate negative numbers: 0 – 2 to –2, 6 – 9 to –3, and so on. With each such identification, the immobilized numbers separate, their subtraction game having found a number to culminate in. By the end, you're quite exhausted—it's as if you've freed an endless colony of birds mired in an oil spill.

How do you feel about your new offspring? Historically, negative numbers were shunned for a long time by humans. Even the math greats like Pascal and Descartes viewed them with suspicion, calling them "false," "absurd," and worse. Mathematicians performed acrobatic calculations to detour around possible encounters with negatives. Textbooks renounced their use as late as in the eighteenth century.

The problem is that while it's easy to see *one* apple or *two* oranges—and demonstrate these tangibly to tots in a math class, that's not possible with *minus one* apple or *minus two* oranges. I once spent a transcontinental flight playing Freud to the passenger next to me who professed severe mathphobia—we finally traced her estrangement back to childhood trust issues with negatives.

Such issues shouldn't arise the way you've just created the numbers. Not only because your universe has no apples or oranges to count or "dis-count" yet, but also because your numbers are independent entities, *all* of them. Minus One has a defining set like any other number—admittedly more complicated than those for the positives, but a defining set all the same. You don't have cause to discriminate.

Lili might even attach a personality to Minus One if she channeled her synesthetic urges. Perhaps she'd make him sullen and resentful because of all the suspicion humans have heaped on him.

Let's come up with a new collective name, *integers*, to recognize the equal validity of the naturals and their negatives. These integers have a greater claim to perfection than the naturals, since one can not only add and multiply among any of them but also play the subtraction game. We can think of them as self-contained under +, ×, and −.

Will these be perfect? Unfortunately, no, as hinted by the fact that the Latin word *integer* translates to "whole" in English.

The horrors of division

In the spirit of the subtraction game that undid the effect of addition, you devise a new game, called *division*, to reverse multiplication. The rule is simple enough: since 4 times 2 is 8, "dividing" 8 by 4 would give 2. Similarly, $3 \times 3 = 9$ so $9 \div 3 = 3$.

As usual, the numbers rush to try this new game. But it quickly becomes apparent that only a few select pairs can hope to succeed at playing it. Most attempts, like $7 \div 16$ or $3 \div 9$, simply fail, since there's no integer available to complete the game. The consequences are even more horrific than those with subtraction: numbers attracted by the thrill of this reckless roulette scrunch their defining sets into each other and emerge shockingly maimed.

Since a similar situation arose with subtraction, this time you're more prepared. You realize you again need to create more numbers—to fill the gaps where all these unresolved "quotients" would lie (i.e., the same way you treated unresolved differences in subtraction). You start by creating the number corresponding to $1 \div 2$, which you denote by $\frac{1}{2}$ (also, alternatively, by 1/2). What should be in its defining set? You decide that anytime a number is divided by two times itself, you will equate the resulting quotient to $\frac{1}{2}$. In other words, all such divisions (such as $\frac{2}{4}$, $\frac{3}{6}$, $\frac{-1}{-2}$, and so on) will be in the defining set for $\frac{1}{2}$.

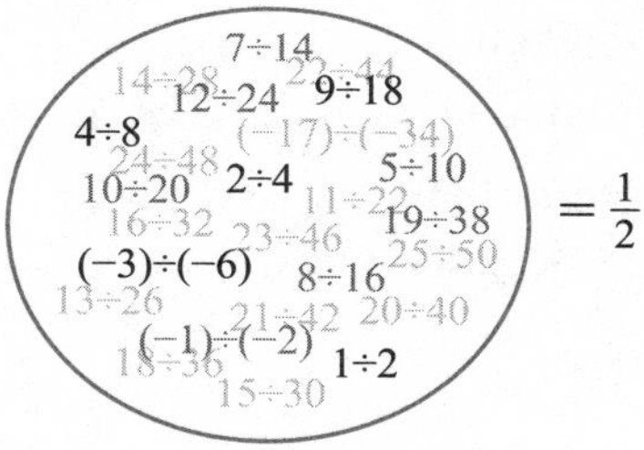

Proceeding this way, you construct similar defining sets for $\frac{1}{3}$, $\frac{2}{3}$, and so on. In fact, you address every division that could possibly be carried out, that is, every single integer divided by every other. The only exception is that you never divide by zero, because you don't know what the division will give.*

Once you're done, you christen these new numbers *fractions*.† With these, you can free all the stuck quotients and restore your universe.

You notice several neat things about your latest infants. First, any integer, say 3 or −3, can also be written as a fraction—simply by

* For instance, what would $0 \div 0$ be? You know $0 \times 1 = 0$, so by the rules of your game, $0 \div 0$ should be 1. However, you also know $0 \times 2 = 0$, which means $0 \div 0$ should also equal 2. The answer can't be both, which is why $0 \div 0$ can't be defined. We'll return to the question of dividing by zero when we talk about infinity on Day 6.

† We mentioned that integers correspond to whole numbers (the Latin *integer* means "whole"). The word "fraction" derives from the Latin *fractio*, which means "to break." Indeed, fractions were originally called "broken numbers."

dividing it by 1 (as in $\frac{3}{1}$ or $\frac{-3}{1}$). Next, these fractions can play all the games you've devised so far—pairs of them can be added, multiplied, subtracted, or divided (as long as zero isn't a divisor). The result always turns out to be another fraction! This shows how far you've come in terms of the perfection of the numbers you've created. Whereas the naturals were only self-contained under addition and multiplication, and the integers only under subtraction as well, these new fractions are self-contained under all four games.

Do you feel buoyed by this? A sense of rightness or satisfaction at things fitting into place? If so, you're experiencing aesthetic appreciation of mathematics.

You start calling the fractions *rational* numbers. This is because they just express the *ratio* between two integers. Your universe, at least for the time being, is orderly again.

Picturing numbers as defining sets

Unless you're a mathematician, the defining sets for negatives and fractions surely must have come as a surprise. Who would have guessed such complexity lay behind innocuous symbols like −1 and $\frac{1}{2}$? Weren't there easier paths you encountered in school that led you to these numbers?

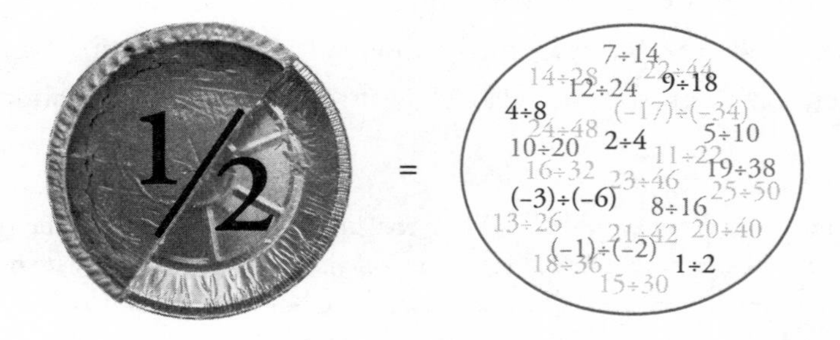

It's true that more applied settings are used in classrooms—dividing a pie in two to formulate $\frac{1}{2}$, for instance. However, if you're building your (as yet pie-less) universe from scratch, there's no option but to define such numbers through first principles. The advantage is that this gives you a glimpse into what lies inside numbers—for instance, that $\frac{1}{2}$ represents not just $1 \div 2$ but also $2 \div 4$, $3 \div 6$, and so on. It's like splitting the atom and finding a cluster of elementary particles inside.

Fortunately, you don't need to keep the mathematical innards of a number like -1 or $\frac{1}{2}$ in mind while using it. The important thing is to have made the exploratory journey at least once.*

There's a deeper reason behind our deep dive into defining sets: to usher you into the world of abstraction, which we'll need for building almost everything in the universe. Although abstraction might make some readers wary, notice that we all use it routinely in our day-to-day lives. For instance, we look at patterns of one-ness and two-ness and three-ness in collections of objects (apples, oranges, whatever) and distill these into the abstract concept of number. Essential qualities like symmetry and beauty are also abstracted in large part from patterns observed through our experience.

The point of abstraction is to strip away unnecessary detail in order to get to the pattern or idea underneath. This is what happens when you ignore whether you're looking at apples or oranges, and note that there are *two* of them. Similarly, when you disregard the shape or thickness of a set's boundary and focus instead on what rule of belonging is being conveyed. With abstraction, you can quickly detect patterns, extend them, and even connect seemingly unrelated phenomena.

Defining sets nicely illustrate these advantages of abstraction. Once you understand the defining sets for Zero, One, and Two, you can immediately (as we saw) extend the abstract pattern to create the defining set for any natural number. And when you encounter

---

* This is similar to comprehending that 6×10 is just 6 added to itself 10 times. It's important to have a secure understanding of such fundamental points, but just as important to be able to file them away afterward and move on (for instance, by learning multiplication techniques that are more efficient than repeated addition).

"stuck" divisions like $1 \div 2$, $2 \div 5$, and the rest, you're able to connect them with the stuck subtractions $1 - 2$, $2 - 5$, and so on, from before. This matching of an abstract pattern leads you to a similar solution: gather together sets of stuck pairs to construct new numbers!

One of the most abstract concepts we'll need is infinity—notice how defining sets have already made you quite comfortable with using it. You first encountered this concept in the Big Bang of Numbers, when you lit the chain reaction that created an infinite succession of naturals. More recently, it popped up when you gathered together infinite pairs of stuck-together numbers to form new defining sets. Most religious accounts of creation involve the infinite, as does physics, where the universe emerges from an infinitely dense initial state. So if you ever felt daunted competing with God or physics, realize that you're now just as empowered as they are in terms of harnessing infinity and using it to create.

One final note. Perhaps defining sets have been too involved—like too much heat in a curry ordered as an act of daredevilry. In particular, slicing numbers open to examine their guts may have made you queasy. (Should we add numbers to the things that, like sausage, you should never see made?) If so, take heart. Defining sets have served their purpose both by making numbers more manifest for us and by opening us up to abstraction. We won't need to invoke them again for any new numbers we create.

4.

SEARCHING FOR THEIR ROOTS

WHILE THEY DIDN'T PROPOSE TO ACTUALLY *BUILD* the universe as you're doing, the Greeks were the first to suggest it might be based on math. "All is number," Pythagoras postulated more than twenty-five hundred years ago. He believed that the laws governing all physical phenomena, from the harmonious overtones of musical instruments to the motion of planets and stars in the sky (*Musica Universalis*, the great symphony of the universe), could be expressed in terms of the ratios between integers.

Such whole number ratios are, of course, precisely the rationals. So, were Pythagoras your guru, the fractions you just constructed would have been the culmination of all the numbers needed.

I'm here to declare Pythagoras wrong. Building the universe takes more than just rationals. You nod wisely at this, since there's a very conspicuous number you've probably noticed yourself we're missing: pi. You recall that dividing the circumference by the diameter for any circle always equals pi, and this value cannot be expressed as the ratio of two integers. There might have been a time in school when you were instructed to take pi's value to be $\frac{22}{7}$, but you later learned this was only an approximation.

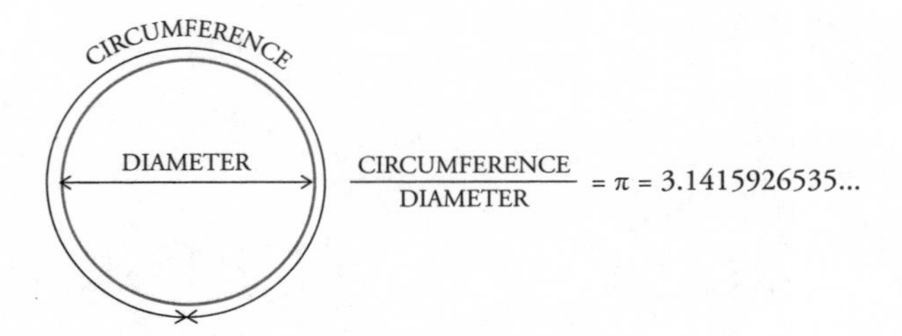

Indeed, pi points to the existence of numbers that are not rational. However, we haven't created geometry yet, so can hardly invoke circles to introduce pi in our thought experiment. Our universe is not just pie-less but also pi-less. Shame on me for including the above diagram, a clear infraction of the rules!*

Consequently, we'll need to motivate such nonfractions through another game. One that's intriguingly different from operations like +, −, ×, ÷, which we usually associate with arithmetic. There's also a surprise in store to further whet your curiosity. The primary benefit to our universe from the numbers this game leads to is not what you—or most mathematicians—might think.

The fifth game

Unlike previous games, the numbers don't need a partner to play this one, so it won't entangle their defining sets. I'll call it the "square root" game, quite indefensibly, since squares—like circles—haven't been created yet. The purpose of the game is to undo the act of multiplying a number by itself. Since $2 \times 2 = 4$, the game takes input 4 and outputs 2, which we define to be 4's *square root*. Since $3 \times 3 = 9$, input 9 gives output 3 (i.e., 3 is 9's square root).†

Numbers, restless as always, dive into this new diversion. Zero

* Actually, there *is* a way to introduce pi at this stage of our thought experiment, using just what we've developed so far. With it, talking about pi here does not violate our rules. See the endnotes.

† There's a negative version of the game where the square roots are −2 and −3, but we'll stick here to the positive.

and Uno are amused to discover that their square roots are just their own selves! Two, however, runs into a problem. At first, he thinks his square root might be the fraction $\frac{3}{2}$, but $\frac{3}{2} \times \frac{3}{2} = \frac{9}{4}$, which is bigger than he is. Next, Two tries $\frac{7}{5}$, but $\frac{7}{5} \times \frac{7}{5} = \frac{49}{25}$, while he needs the answer to be $\frac{50}{25}$ for it to equal himself. Even a fraction as fine-tuned as $\frac{141}{100}$ doesn't quite work, giving the square $\frac{19881}{10000}$, which is slightly too small. Two never finds his exact square root, no matter which candidate he tries.

Perhaps you urge Two to continue his quest (as Pythagoras might have done). Surely Two hasn't combed through the entire collection of rationals yet. But then Three encounters a similar problem, and Five shows up with the same complaint. Zero finally breaks the news to you—she's meditated on the problem enough to determine no fractions can fit the bill.

For a while you wonder if you could just declare such missing square roots—those for 2, 3, 5, for instance—to be new numbers. After all, isn't that what you did to create the negatives and the fractions? You've even designed a cute hat-like symbol for these new creations—you'll denote them by $\sqrt{2}$, $\sqrt{3}$, $\sqrt{5}$, and so on. But then you hear about something called the cube root game, which gives you pause.

Apparently, some numbers, like 8 and 27, unable to split into square roots, have invented their own variation. The idea is similar: since $2 \times 2 \times 2$ equals 8, they've declared that the *cube root* of 8 equals 2. Similarly, the cube root of 27 is 3. The problem is that many numbers can't play this game either, leading to a new litany of complaints.

It gets more chaotic as other disgruntled numbers start inventing their own games. For instance, 32 creates the fifth root game, through which she declares that her *fifth* root is 2. Then 128 comes up with the seventh root game in a similar vein, and you even hear of tenth and twentieth roots. How are you supposed to beget all these strange new numbers when you're not even sure you can dream up symbols for them?

Soon, every number is protesting that it can't play some game or other. Just when it looks like your universe is going to degenerate into unruliness comes a revolutionizing development. The numbers discover decimal expansions, an alternate form they can take.

The double life of fractions

One of the perks of being in a thought experiment is that we don't have to labor through each detail. Decimal expansions stem from the process of long division—something I (and I'm guessing you) remember being a slog to learn as a kid. There's no need to bring up dividing 1 by 2 to get 0.5 or 1 by 3 to get 0.3333 . . . again. Let's instead just say the numbers make such discoveries through their endless games. Perhaps $\frac{1}{10}$ finds out first, realizing one day that he can transform himself into 0.1. Probably $\frac{1}{100}$ changes to 0.01 soon after, followed by $\frac{1}{1000}$ to 0.001. Eventually, every fraction gets the hang of it. One incentive that keeps them going is to be able to play the greater-than/less-than game that naturals play. While it's hard to figure out whether (say) the fraction $\frac{2}{5}$ or $\frac{3}{8}$ is greater, this becomes apparent when they convert themselves to their respective decimal forms 0.4 and 0.375. They can now use the ordering of the naturals, where 4 is greater than 3.

Whole numbers experience a change as well, starting with Zero, who realizes she can express herself not just with a single 0 but with an endless train of them,

$$0 = 0.0000 \ldots.$$

Her vision resolves a question the fractions have been wondering about—why some of them, like $\frac{1}{2} = 0.5$, have a finite decimal expansion, while others, like $\frac{1}{3} = 0.3333 \ldots$, go on forever. Zero's discovery shows that *all* numbers go on forever—they just need to add a train of zeros after their last digit, as in

$$\frac{1}{2} = 0.50000 \ldots.$$

As a result, numbers start appearing as long rippling strings of digits, like eels with exposed skeletons. The new X-ray vision you seem to have developed allows you to gaze right through their skin.

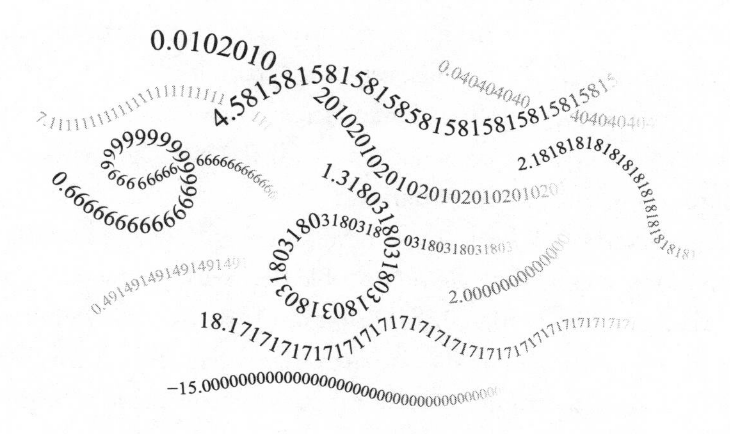

Watching an elegant decimal undulate by in slow motion, its elongated body consisting entirely of glinting sevens, a question occurs to you. Might every rational end in some repeating train of digits? For instance, 2.0000 . . . and 4.3333 . . . both end in a single repeating digit. Some numbers, like 6.212121 . . . , have two digits that repeat, while 3.581581581 . . . , you notice, has three. Others have more, like the innocuous-seeming $\frac{11}{7}$ that floats by, surprising you with the six repeating digits in its expansion of 1.571428571428 . . . .

Ruminating on this, you realize your hunch is correct. No matter whether a number is whole or fractional, its decimal expansion will always end in some string of digits repeating indefinitely.*

---

* To see this, consider long division by 7, for instance. Then at each step, you can have only 7 possible remainders: 0, 1, 2, 3, 4, 5, or 6. If you ever encounter remainder 0, the division is exact, and you're done. Otherwise, you can expect a different remainder for at most 6 steps before encountering a repetition, i.e., 1, 2, 3, 4, 5, or 6 again. Once you get the same remainder, the digits in the quotients will also, of necessity, start repeating. Try a few examples like 3/7, 25/7, etc., to get a feel for why this has to be true. Then try long division by some other denominators, say 9 and 11.

The nonrepeaters

The observation about rationals always having a repeating expansion sparks an idea. What if you were to remove this constraint, and let the digits unfurl in any and every way possible? Wouldn't that produce new numbers? If so, perhaps those with expansions that *don't* repeat are exactly the ones needed to fill the holes unearthed by the square root game and its variations!

You decide you'll build these new numbers. It's hard to isolate just them alone, so you'll create *all* possible decimal expansions, repeating and nonrepeating (which, in addition to giving the new numbers, will also give the rationals again, as part of the mix). This turns out to be quite a task. For instance, for numbers whose expansion begins with 4.1 . . . , the next digit can be anything from 0 to 9, that is, you have to include all numbers starting with 4.10 . . . to 4.19 . . . . Then, for each of these, the digit that follows has to again be arbitrary, as does the digit after that, and the one after, and so on. And you have to do the same with numbers beginning with 40.1 . . . or 400.1 . . . or 4000.1 . . . . In other words, you have to string together infinite numbers of infinite expansions. Creating just the rationals felt like a breeze in comparison.

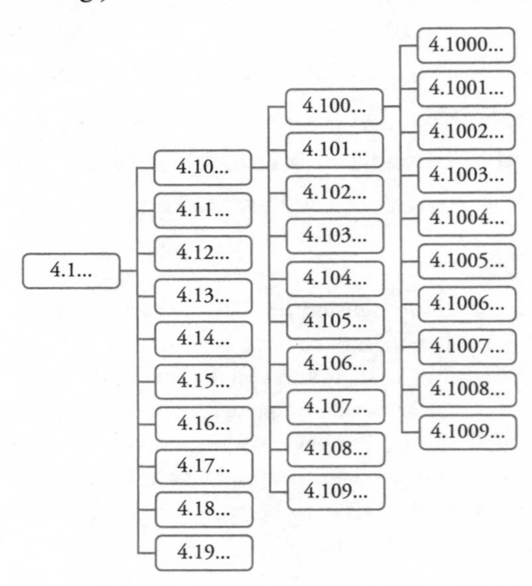

Perhaps you'd never have embarked on this venture had you known you'd have to work so hard. Thankfully, Uno sells it to other numbers as a game, mobilizing an army of them to assist.

The task finally gets completed. The rationals cheer the new siblings in their midst. Now comes the moment of truth. There's no guarantee that all the numbers lacking roots will be able to find them in your new creations. But then you see Two triumphantly point to a long decimal that starts with 1.4142 . . . and figure this must be his square root. Other numbers, too, come dancing to you arm in arm with the roots that had eluded them. Curiously, only some of the new numbers you've built are claimed as roots. The vast majority seem completely independent, as if unrelated to any rational. Pi, which I slipped in a mention of earlier, happens to be one of them.

The new numbers, the ones with decimal expansions that don't repeat, start calling themselves *irrationals*, to distinguish themselves from rationals. You coin a new word, *reals*,* to encompass both the rationals and the irrationals. Watching this family of reals mingle together gives you a warm glow of parental accomplishment.

---

* The name comes from René Descartes (of "I think, therefore I am" fame), who called them "real" to distinguish them from "imaginary" numbers, which we'll encounter shortly.

5.

AN IRRATIONAL UNIVERSE

SO HOW DO THE IRRATIONALS YOU'VE JUST CREATED enhance your universe?

On an aesthetic level, they provide a sense of completeness. You can rest secure in the knowledge that you've created each and every decimal expansion possible, without any omissions. If you felt a touch of mathematical appreciation producing the rationals, that's now grown to full-fledged intellectual satisfaction. However, this benefit is somewhat abstract. What do irrationals contribute in *practical* terms to your universe?

Most mathematicians would answer, *precision*. With irrationals at your disposal, you can identify the exact square root of 2, the exact ratio of circumference to diameter. But there's a gap here, since you can never actually write down the exact value of $\sqrt{2}$ or pi. Each has an infinite expansion, so in reality, you are forced to use an approximation.

And that's fine. For instance, NASA has been reported to use the value of pi to fifteen decimal places, $\pi = 3.141592653589793$, for all its calculations. This turns out to be precise enough to track even its most far-flung satellites. As one of NASA's mission managers has stated, were they to use forty digits instead, they'd be able to calculate the circumference of the entire visible universe to an accuracy of the diameter of a hydrogen atom! In other words, it would be overkill.

NASA's experience suggests that while building our universe, you might never need the exact value of an irrational—you could get away with approximations that were close enough. In other words, rationals, with enough decimals, would suffice. Which returns us to the question, what *do* irrationals bring to the table?

Later, we'll see how essential irrationals are for motivating the construction of a straight line (without which, we'd be unable to create the 3-D space we live in). But right now, let me introduce another crucial aspect they inspire in the universe. To do so, I need to first take you on a detour, by acquainting you with a very curious irrational many of you are unlikely to have seen before.

A crystal ball of digits

Imagine a decimal point with each of the whole numbers 1, 2, 3, . . . strung together after it, in order:

0.1 2 3 4 5 6 7 8 9 10 11 12 13 14 15 . . . .

I've written down the first fifteen naturals above, so the next two digits would be 16, the two after that 17, and so on. The complete expansion 0.12345 . . . would contain the sequence of *all* the naturals. Notice this gives a real number lying between 0 and 1.

Clearly, this infinite expansion doesn't repeat—so the number can't be a fraction, it must be irrational. Let's call it the "crystal ball" irrational.* Why? Because it contains the answer to every question about the future posed by you or anyone else!

To see this, think of the simple numerical codes used by kids in their cloak-and-dagger games. I remember making the substitutions a = 1, b = 2, c = 3, up to z = 26. (Perhaps you tried this as well?) Using this code, the word "cab" becomes the whole number 312. One can convert sentences, articles, even books this way—after adding some extra integers for punctuation, spaces between letters, and so

* The number's mathematical name is the *Champernowne constant.*

on. For instance, the entire 587,287-word text of *War and Peace* can be reduced to a single integer using the code! A very long integer, it's true, but still one of finite length.

Now, the crystal ball irrational 0.12345 . . . contains *every* natural number, so it must contain the one corresponding to "cab." This is easy to verify—just read along the list to the 312th natural listed, and voilà! It's there.

But by the same token, this irrational also contains the much larger integer for *War and Peace*. Proceed along the list far enough and you'll eventually have to get to it. In fact, you can find the entire Harry Potter series in this irrational number, along with all the plays by William Shakespeare. Everything ever written—every poem, every treatise, every issue of the *New York Times* exists in it. Moreover, it also contains everything that ever *will* be written in the future. Tomorrow's newspaper, for instance—or one from next year. Which means that if you're wondering what stocks to buy or whether we'll be submerged due to melting ice caps, the answer's in there.

This book I'm writing is in there as well—wouldn't it be great if I could just copy it down without having to actually slog through all the work ahead? Unfortunately, it's inaccessible, since I don't know at which digit it starts. I also have no way of differentiating my book from all the slightly changed variations contained in the number, or from radically altered versions, or, for that matter, from the many tracts of utter gibberish. The same is true of what stocks to purchase, since mixed in with what's actually going to happen are reports of every possible future imaginable.

In 1941, the Argentine author Borges published a short story called "The Library of Babel" about a rambling multiroomed library that holds every book of 410 pages ever written or that ever might be written. However, nobody knows where anything is shelved, so the entire collection is useless. The crystal ball irrational is an expanded version of that library, since it stores every bit of information that might ever exist. A virtual flash drive if you will, which, sadly for us—or perhaps fortunately—is perfectly encrypted.

More flash drives

You might think the number we've discussed is unique in terms of its crystal ball potential, but it's not. Mathematicians have established that almost every irrational number has the same property of containing every possible string of digits somewhere in its expansion. But there isn't a guide to pinpoint where any string begins. This includes numbers like $\sqrt{2}$ and pi, it's suspected.

What evidence is there that pi might have this property? Certainly it's not hard to find the word "cab" in it, that is, the string 312. This occurs starting at position 2631 after the decimal point, then again starting at position 3321 (in fact, the string 312 occurs 1,999,464 times in the first 2 billion digits!).*

But how about *War and Peace*? How do we know the enormous integer corresponding to it occurs somewhere in pi? Although mathematicians have calculated pi to trillions of digits, nobody, to my knowledge, has located Tolstoy's doorstop in it.

What mathematicians *have* verified through experiment, however, is that the digits 0 through 9 occur with about equal frequency. For instance, here is the proportion in which each appears in the first trillion digits of pi:

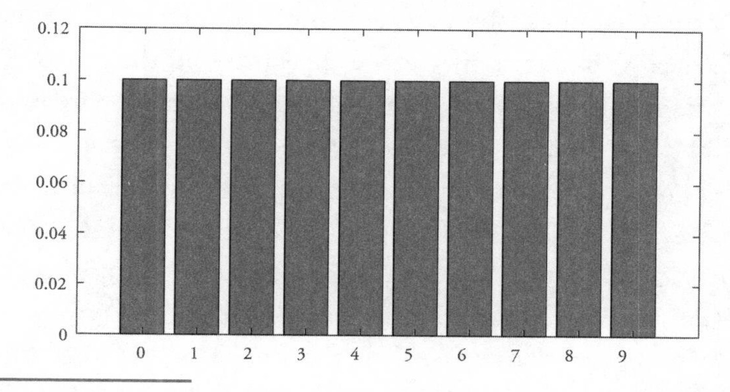

* You can try to locate any numerical string (e.g., your birthdate) in the expansion of pi on the website http://pisearch.org/. This website also lets you search for strings in the expansion of $\sqrt{2}$.

Zooming in to the top of the bars, one sees a little variation, but each digit still has about 10 percent frequency.

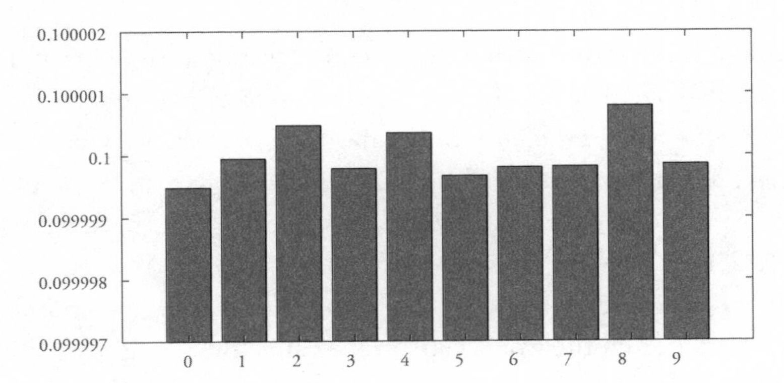

Similar experiments have also verified that each two-digit string (e.g., 13 or 27 or 68—there are a hundred of these) appears with near-equal frequency, which works out to about one-hundredth, or 1 percent, of the time. Moreover, three-digit strings like 312 each appear approximately one-thousandth, or 0.1 percent, of the time (the previously stated tally of 1,999,464 occurrences in 2 billion digits indeed translates to about 0.1 percent). Continuing this way, we can expect every four-digit string to occur 0.01 percent of the time, every five-digit string to occur 0.001 percent of the time, and so on. Eventually, we reach strings with length equal to the one for *War and Peace*. All these strings again can be expected to occur with equal frequency. Which means that not only can we expect to find the code for *War and Peace* in the expansion of pi but that it should occur multiple times, and just about as often as any other string of that length. In addition, most other irrationals share this property (for instance, the string 312 occurs 2,002,388 times in the first 2 billion digits of $\sqrt{2}$, which is again about 0.1 percent). Consequently, we can expect to find *War and Peace* in their expansions as well.

So yes, the irrationals have all the knowledge of the universe coded and locked up within them. But what does this have to do with building the universe? The knowledge is inaccessible, so how does it help?

Randomness

OK, all this was the detour—or subterfuge, even—before I unveiled the irrationals' most important contribution. The answer: it's not the coded knowledge they contain, but rather the way they scramble that knowledge! Specifically, it's the *randomness* with which digits appear in their expansions. Randomness will be essential for the construction of your universe, perhaps for generating life itself, as we'll see on the last Day of creation in this book.

Let me qualify this a shade. True randomness is hard to come by—the generally accepted wisdom is that one needs a physical process to generate it. This might conjure up images of dice being rolled or coins flipped, but such outcomes are entirely determined using the laws of physics (and could be calculated if you had the knowledge and patience). So strictly speaking, they don't qualify.* Instead, one needs to track more esoteric phenomena from nature. For instance, it is well known that radioactive elements lose their radioactivity over time by decaying into simpler elements and emitting radiation in the process. The time it takes between successive atoms to decay is unpredictable, so one could measure such gaps (i.e., gaps between spikes in radiation) and use the sequence of times recorded as the basis of generating random numbers. Even this gets rather complicated—hinging on delicate theoretical questions like what constitutes "true" randomness and what physical assumptions are being made.

While physicists will argue its fine points, persons of faith might reject this nature-based notion of randomness entirely. Ascribing everything to an omniscient creator implies, ipso facto, that radioactive decay also follows *the Plan*. "Not a leaf falls but that He knows it," says the Koran, a sentiment that logically must extend to the ejec-

---

* Neither do processes involving human choice. As an example, I may try to select a sample of students "randomly" from my class, but some conscious or unconscious bias will always creep in.

tion of electrons from atomic nuclei as well (the Bible and Vedas contain comparable sayings). Under such a worldview, although things might appear unpredictable, nothing can be truly random. So this seemingly innocuous issue devolves quickly along familiar fault lines. Randomness versus God—can only one exist?

Fortunately, given your vantage point that predates the universe, you don't have to referee this. The match hasn't begun, neither physics nor God has stepped into the ring yet. Since there have been no physical processes so far, you can hardly define randomness based on them. The only understanding—nay, the only *inkling*—you can hope to have of randomness (given what the universe currently contains) comes from the irrationals.

Indeed, as you examine $\sqrt{2} = 1.41421356237\ldots$ (for instance) you realize it has none of the order you saw in the decimal expansion of fractions. The train of digits swirling by now seems to change and shift every instant. You can't readily discern any pattern in how the expansion unspools, how one number follows from the other. You call the sequence *random*—to capture its variation, its unpredictability, its arbitrariness.*

But there's something rather compromised about this randomness. That's because the sequence is entirely predictable—the key you need to generate it is $\sqrt{2}$ itself. The digits only look unconnected—in reality, they're simply following the dictates of the number's expansion. There's no arbitrariness allowed—change a single digit and you'll have changed $\sqrt{2}$ to something else.

This type of "pseudo" randomness is the best one can expect from irrationals. But it's good enough for our purposes, sufficient to model the variety and serendipity we want for your universe. In many settings, the numbers generated from irrationals are essentially indistinguishable from sequences that are "truly" random. Whether

* Most, but not all, irrationals will exhibit such randomness. An obvious exception is the crystal ball irrational.

such idealized "true" randomness even exists is a question we'll leave for another time.

For now, note how the irrationals provide a compromise between the conflicting worldviews of science and theism. On the one hand, irrationals exist independently of any external agency, so the randomness derived from them owes its provenance only to our starting point, the humble empty set. On the other hand, they yield only a pseudorandomness at best, every aspect of which is governed by the number itself. It's an apt metaphor for a presiding entity silently exerting control over everything, even if it's through randomness.*

Let me end this chapter on a semantic note. Do you feel the name "irrational" connotes something negative? An absence of rationality, of logic, of common sense? I used to think the choice unfortunate. Why taint an entire class of numbers like that?

But now I've begun to feel the name is affirming. Irrational numbers introduce a contrarian element, one that rebels against white-bread orderliness. An irrational universe is one that possesses a touch of whimsy, of mischief. A universe that's more interesting, because it has incorporated randomness.

---

* The 1998 thriller movie *Pi* cleverly uses this idea as its premise, positing that everything in the world, from messianic messages in the Torah to the behavior of the stock market, is governed by the decimal expansion of the number pi.

6.

JOURNEY INTO THE IMAGINARY

BY NOW YOU'VE PROBABLY REALIZED HOW EXHAUSTING the act of creation is, even if one is only formulating concepts. After six days of Genesis, even God takes a rest. Having constructed every decimal expansion possible, surely you deserve a break yourself. The reals you've created are seamlessly complete, achieving a peak of mathematical perfection that towers over your previous attempts. In this wonderland of yours, each number can play every game ever invented to its heart's content.

Alas, that's not quite true. Your Eden, it turns out, is still deficient.

Rooting for negatives

It's the fifth game, the one of taking square roots, that continues to cause inequity. Through their games, the numbers have seen that every number multiplied by itself always gives a product that's zero or positive, which means the square of any number can never be negative. This implies there can be no real number that equals the square root of –1, or that of any other negative. So the negatives are still excluded from playing the square root game.

But is this really an issue? Weren't you taught early on in school that the square root of a negative number does not exist? What would something as bizarre as $\sqrt{-1}$ correspond to, anyway? You know you're not supposed to let everyday experience weigh in too heavily in this thought experiment, but let's face it—so far, it's been a big help. Constructing the negatives, the fractions, the irrationals has all gone smoothly because your prior exposure to −1, $\frac{1}{2}$, and pi made such numbers easy to accept. Now, however, your intuition fails you. You've seen $\frac{1}{2}$ an apple, bitten into several yourself. But how to wrap your head around $\sqrt{-1}$ of a piece of fruit?

The short answer is you can't. In fact, you shouldn't even try, because doing so means falling into the ever-waiting trap of associating numbers only with quantity or counting. That's what we've all been conditioned to believe through our years of schooling—that numbers are inextricably tied to objects in need of enumeration, that their identity starts and ends with such connections. Despite the lessons of the previous chapters, it can still be difficult, deep down, to regard numbers as truly independent entities.

But with $\sqrt{-1}$, there's going to be no choice. Its existence cannot be motivated by pointing to uses that are commonly known. Rather, the only way to proceed is to put one's full faith in the premise that all numbers, like people, are created equal. Accept that, and you'll have to agree that since every positive has a square root, so should every negative.* Therefore, there must be an empty slot with designation $\sqrt{-1}$ waiting to be filled. It's like knowing to search for a new chemical element based on an unoccupied position in the periodic table.

Perhaps this argument smacks too much of aesthetics and mathematical completeness, which you may not have fully warmed to as yet. Your reals, the combination of rationals and irrationals, are already a crowning achievement. So what if they have this single shortcoming of missing square roots for negatives? Couldn't you just appreciate the reals as they are, learn to ignore this problem?

You could. But that would counter the aim of this book to

---

* If this sounds weird or unnatural, put it down to personal prejudice, which you need to overcome. "Who am I to judge?"—remember the disavowal the pope once made.

broaden horizons. Every question that arises opens up an avenue for exploration.

And this particular avenue will prove essential. The number $\sqrt{-1}$ will play an unexpected, but indispensable, role in powering your building project's geometry. Without it, your universe would remain, quite literally, one-dimensional.

Minus One gets a sister

So how to create $\sqrt{-1}$? How to split -1 into identical square root components? It's pointless searching among decimal expansions, since they've all been used up to fashion the reals. In fact, you already know $\sqrt{-1}$ can't be real, so you need to look outside this set. But nothing exists outside the reals, so you have to create $\sqrt{-1}$ from scratch.

You've done this before, so don't waver. Remember how you created -1 to be a new number lying outside the naturals, $\frac{1}{2}$ a new number outside the integers, $\sqrt{2}$ outside the rationals? In the same spirit, define $\sqrt{-1}$ to be a new number, as elemental as 0 or 1, which lies outside the reals. That's all there is to it. The key is not what goes into the slot for $\sqrt{-1}$, but where it's located.

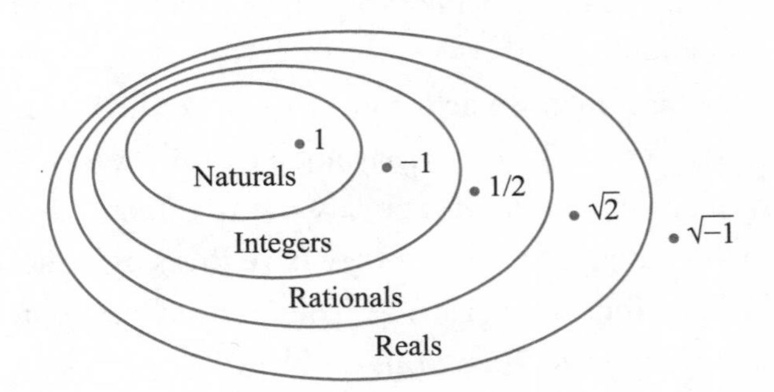

You might have been expecting something more technical, but recall that to simplify things, we've dispensed with defining sets. Perhaps the construction still seems too pat. To reassure you that you have an actual number, let's look at it through the eyes of Minus One.

Someone who's borne the brunt of discrimination against the negatives, who's bristled at being a second-class citizen, unable to play the square root game. As he gazes at his tiny new sister $\sqrt{-1}$, Minus One's resentment starts to fade. He hugs her close to himself and vows to protect her in every way. "I'll help you build a new universe of numbers just like yourself, somewhere far away."

Minus One calls his sister Ima, which we'll abbreviate to i (so that $i = \sqrt{-1}$). He begins to play with Ima, teaching her all the games he knows. The very first game, where Ima multiplies herself, shows how she and her brother are linked. This game gives the equation

$$i \times i = \sqrt{-1} \times \sqrt{-1} = -1,$$

which demonstrates that Ima is exactly a square root for Minus One!

It's the next game whose outcome will be so important for your universe. Ima starts multiplying reals by herself, to form products like $2i$, $-i$, and $\frac{i}{2}$. Each of these is another number, so she ends up creating a huge new class, which are all multiples of i. To emphasize their independence from the *reals*, these numbers adopt a diametrically opposite name, calling themselves the *imaginaries*. Or as i notes, *Ima*ginaries.

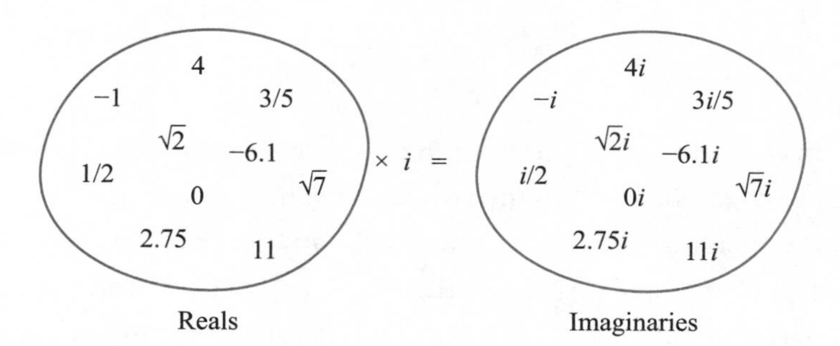

The reals are a bit reserved at first, but once the ice breaks, they start eagerly engaging with their imaginary siblings. It's true what they say about opposites attracting, because addition between the two groups becomes the hottest new game. Sums appear everywhere: $2 - 3i$,

$-4 + 8i$, $\frac{1}{3} + \frac{6}{5}i$, every single real added to every single imaginary, in one big, jumbled assemblage. Such hybrid numbers start calling themselves *complex*.

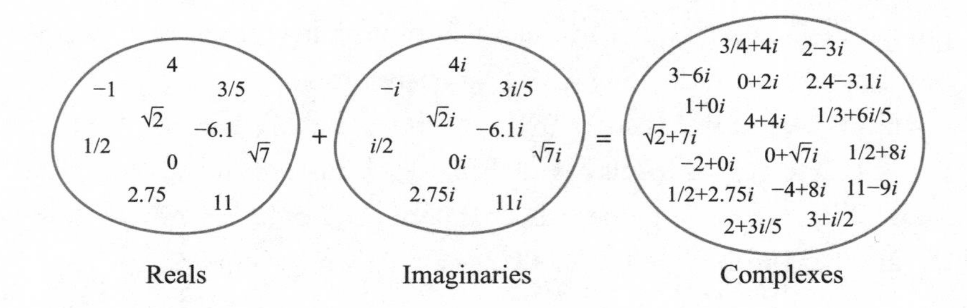

Notice that the complexes include all the reals, since any real number, say 3, can be expressed as

$$3 = 3 + 0i$$

showing it is the sum of a real (in this case, 3) and an imaginary ($0i$, which is just 0). Similarly, the complexes also include all the imaginaries, since any imaginary number, like $2i$, can be expressed as the sum of a real and an imaginary as well:

$$2i = 0 + 2i.$$

You may have a hard time thinking of complexes as single numbers, since they obviously combine two separate quantities—the real part and the imaginary part. But notice that any real number (say pi) is the sum of two parts as well—the whole part (i.e., 3) and the decimal part (i.e., 0.14159 . . .). So is any fraction, which incorporates a numerator and a denominator. This shows you've already had some experience with such composites. It's this very two-ness that will make complexes useful for us, as we shall see.

In terms of number systems, the complexes, finally, are *it*, the holy grail! The *ne plus ultra* collection in which numbers can not only add, subtract, multiply, and divide among themselves, but even take arbitrary roots.* We've reached the end of an evolutionary chain, entered the utopia of a perfect number universe.

The issue of practicality

Here's an obvious question. If complexes are so perfect, so evolved, why haven't we seen more of them in our lives? What are they good for, anyway?

Let's make a distinction right away—there's a big difference between how "evolved" a number system is and how *useful* it is. The hierarchy we've developed is geared toward performing an expanding list of mathematical operations, resulting in increasingly sophisticated number systems. This sophistication is best appreciated intellectually, aesthetically, rather than in terms of only practical applications.

In terms of frequency of use, it's hard to beat the naturals. They're always going to be the most indispensable, the most ubiquitous. By the time you get to the complexes, the applications are more rarefied.

---

* Notice that the imaginaries, by themselves, are not self-contained under multiplication, since the product of two imaginaries is not an imaginary, but a real (e.g., $i \times i = -1$). One has to take the sum of the reals and imaginaries, i.e., the complexes, to get a set that's self-contained—under not just multiplication but also all the other operations we've considered. For more, see the endnotes.

You're unlikely to encounter them except in technical areas of science and engineering.*

And yet, they turn out to be indispensable while building up the universe brick by brick from the numbers, as you'll see on Day 2 (they'll also prove their worth on Day 4). That's the beauty of the narrative we're following. It's going to give you an appreciation of the complexes in practical terms as well.

Speaking of practicality, here's another question. What to make of the nomenclature "imaginary"? Does it make it easier to dismiss such numbers because they're mere figments? Why should anyone take them seriously?

The ironic thing is that *all* numbers are figments. Or, more respectably, constructs. The imaginaries might not be applicable in counting or sizing, but they're no more nor less "real" than the reals.

If we can keep this clarification in mind, "imaginary" is not such a bad name. It points to the role *imagination* plays in mathematics, to the fact that we can build so much structure out of pure concept. Real numbers (which include the naturals) do, indeed, end up being used far more—so in that sense, "real" could be identified with the more practically applied aspects. Then "complex," which combines real and imaginary, becomes a metaphor for the subject itself. Mathematics arises both from practically useful questions and from allowing the mind to run free.

* For instance, they pop up when a physical phenomenon is described by two quantities rather than one. An example is electromagnetism, where there is both an electric field and a magnetic field—by taking the respective strengths of these fields to be the real part and the imaginary part, this duo can be described by a single complex number.

A table of numbers

As Day 1 turns into night, let's summarize what we've accomplished today. Starting with nothing, we've given birth to the numbers that will be our building blocks. In doing so, we've had help from the games of arithmetic: basic operations like subtraction and division giving rise to the integers and rationals, and the square root game leading to both the irrationals and the imaginaries. We'll frequently refer to these different kinds of numbers in the Days to follow. To look them up, just turn to the table below.

Type of Number	**Description**	**Examples**
Naturals	Positive whole numbers used for counting	$0, 1, 2, \ldots$
Integers	Whole numbers, positive and negative	$0, \pm 1, \pm 2, \ldots$
Rationals (Fractions)	Ratios of whole numbers, expressible as decimals with repeating expansions	$\pm\frac{1}{2}, \pm\frac{4}{7}, 1.2, -0.8333\ldots$
Irrationals	Numbers with nonrepeating decimal expansions	$\pi, \sqrt{2}, -\sqrt{3}, 0.12345\ldots$
Reals	Union of rationals and irrationals	$\frac{4}{7}, -\pi, 1.333\ldots$
Imaginaries	Real numbers multiplied by $i = \sqrt{-1}$	$\sqrt{3}i, -5i, 1.2i$
Complexes	Sums of reals and imaginaries	$\pi + 1.633i, \frac{3}{4} - 7i, 0.1 + \sqrt{2}i$

Day 2

GEOMETRY

Using points and numbers to create space

7.

THE UNIVERSE NEEDS ITS SPACE

NOW THAT YOU HAVE THE NUMBERS YOU'LL NEED, it's time to start thinking of housing them, along with all the future creations your universe will contain. You need to create space, or more generally, geometry.

How to go about this? Space is so enmeshed with our experience that we don't think of it as something that needs to be laid out. Imagine the universe around us emptied of everything physical, and what are we left with? A vast expanse of empty space. But did this space always exist? Not in our thought experiment, it didn't. That's because we started with *nothing*. There was no vacuum, no invisible ether, no empty expanse, no volume waiting to be filled. There was only nothing, which means empty space has to be *created*.

Certainly, God takes space for granted. Creation myths are filled with cosmic eggs hatching, skies separating from oceans, even a magical bird stomping the earth into shape. But there's never any mention of laying down a canvas beforehand that can accommodate such objects—objects that possess length and height and width. A lotus stem curls up from Vishnu's navel and then sprouts Brahma from its petals—presumably, the 3-D space needed for such action to unfold is already in place. The book of Genesis starts with God creating heaven and earth, but gives no account of prior prep work to set up the empty stage.

This is the gap you are being called upon to fill. To build a matrix of points or locations, collectively called space, in which God can orchestrate all the ensuing drama to take place.

Physicists acknowledge the need to account for space, but do not require it to be prefabricated as we do. Rather, it emerges explicitly in the creation process. Contrary to what its name suggests, the big bang is not a fireworks-type detonation in which matter explodes from a source and goes flying off in all directions. Yes, the super-concentrated universe rapidly expands from an infinitely dense singularity, but accompanying this is a dilation of space itself. As physicists will take pains to clarify, the big bang is not an expansion *in* space, but an expansion *of* space. In other words, the underlying matrix that accommodates the cosmos is laid out simultaneously as well.

What this implies is that for physicists, space does not exist before the big bang but arises as a byproduct of the contents of the universe. It exists to express their interactions; one possible reason it might be generated is simply to allow particles to be separate. So you can never have empty space by itself, devoid of all traces of the physical universe. This is in contrast with our thought experiment, where we will create empty space first, and later create the tangible objects it will accommodate.

The advantage of our approach is that it allows us to understand space and its mathematical properties independently, an understanding that gets murky when viewing space as a physical artifact. Is there a recipe to create space? How does the fabric of space emerge from the individual locations being incorporated? Could this process lead to different fabrics, creating alternative universes? These questions, as we'll see, are more about mathematics than about physics.

Our goal will be to create different mathematical models—let's think of them as molds—for space. When the big bang occurs, physical space can then just expand into the appropriate mold. So the two approaches would converge to the same result, eliminating the need to arm-wrestle with physicists.*

The first person to systematically consider geometry was the Greek mathematician Euclid, in around 300 BCE. His resulting work, *The Elements,* has had so many readers it's second only to the Bible in terms of the number of editions published. Even Abraham Lincoln carried a copy in his saddlebag to master its principles of logic, which he believed were helpful in framing legislative arguments. The reason Euclid has been so influential is that he pioneered the idea of breaking a subject into its basic building blocks and then deducing all subsequent results from these fundamentals—a technique that's been used in various other fields, from physics and chemistry to economics and linguistics. Clearly, we're following his "building block" idea as we construct our universe.

However, Euclid's specific building up of geometry doesn't fit our agenda so much. That's because his goal was not to construct the underlying matrix of points (which *The Elements*, like the Bible, tacitly assumes is already in place) but rather, to establish all the theorems about circles, triangles, quadrilaterals, and so on, known back then (for example, the Pythagorean theorem). Consequently, he considered lines, planes, and 3-D space to be givens—basic ingredients whose existence did not have to be justified, and which he could take for granted. In contrast, our aim will be precisely to show how these "givens" of Euclid can be created, since we start off with nothing. Fortunately, some of the fundamentals Euclid laid down to *analyze*

---

* Well, maybe not. There would still be the question of how to account for time—which in physics is inextricably linked with space (so that any mold would have to be for spacetime, not space alone). We'll address this issue on Day 5, when we get to the laws of physics. Remember that our primary objective is to show how math figures in questions about origins, not to formulate an airtight origins theory. So we can give ourselves some leeway, in contrast to the constraints under which physicists must operate.

geometry are also useful in *building* up the underlying points. We'll bring out this connection at the end of this chapter.

Later on during this Day, we'll show how some of Euclid's fundamentals, which we all might believe to be irrefutably true based on our experience as humans, can be altered to yield alternative "curved" geometries just as viable as the usual 2-D planes and 3-D space of "Euclidean" geometry we're familiar with. (If you're worried about too much abstraction, take heart—we'll use a very tangible way to explore this.) This is one of the key examples we'll present in this book of how mathematics challenges our fundamental beliefs. Later on, on Day 5, we'll see how such alternative geometries play a role in forming spacetime for our universe, so they're not just mathematical curiosities.

More than Euclid, though, our guide will be the numbers. Although they can't *create* geometry, they will be the inspiration behind it. They will provide us the reason for it to exist.

"Everything starts from a dot"

The Russian artist Wassily Kandinsky is the perfect patron saint for this Day of geometry. We'd like to eventually populate our universe with the same components he used to build his modernist compositions: lines and curves, squares and circles, triangles and rectangles, and so on. Before we get to these shapes, though, we're going to set down the groundwork by following an essential truth contained in his above quote about the dot. What did he mean by it?

Kandinsky was referring to the process of creating a painting, of transforming an abstract mental idea into physical art. Each such realization has to start with brush meeting canvas—and if you think of it, what marks this first instant of contact is always a dot. In other words, the dot is the source from which art springs, the initiator of the painting to come.

Instead of a painting, what we now have in mind is an unformed idea of "space" that we'd like to create in your universe. To do so, we'll follow the instruction inherent in Kandinsky's quote, but with one essential substitution: "point" for "dot." What's the difference?

A *dot* might be tiny, but still covers an area. That's how it contributes to a painting—by tinting a portion of the surface with its own color. Print and electronic images also use this principle—now the dots are pixels. Indeed, pictures can be composed of dots the same way matter can from elementary particles—so, in this sense, Kandinsky's quote can be appreciated in terms of not just art but also physics.

A *point*, on the other hand, is dimensionless—it possesses neither length nor area nor depth. ("That which has no part" is how Euclid defined it.) All it does is mark position—that's its *raison d'être*. Think of it as an idealized dot—one that has shrunk to be smaller than the smallest speck—what mathematicians call an "infinitesimal." Such points are what we'll use to build up geometry, but the way they work is crucially different from Kandinsky's dots. While physical images and objects can be composed of a *finite* amount of paint or pixels or particles, we'll need an *infinite* number of points for the geometric constructs (such as line segments) to come.

So, to express this idea that's in our minds, let's start by introducing just a single point into your universe. This causes an instant change. Whereas before your universe was entirely undifferentiated, like Kandinsky's blank canvas, there's now a marker that interrupts this void. A single, fixed location that we'll call the *origin*, which pierces through the haze of your numbers' environment. Zero, who's always despaired about the insubstantiality of her surroundings, is elated. She squeezes close to the point, using it to anchor herself.

Where did this point come from? Its appearance has to be assumed—one of those "extra building ingredients" I warned in the introduction might be needed in our thought experiment. We'll talk more about such assumptions at the end of this chapter. For now, let's watch some of the numbers making a game of trying to dislodge the point, but Zero shooing them away. The point remains rooted at its location, with single-minded intent.

Seeing Zero so settled, you realize you need another point—for Uno. You're fine with expanding your list of extra building ingredients to accomplish this. Assume, therefore, that you're able to introduce a duplicate point, to create a second distinct location in your universe. Uno's delighted when you assign it to him. "How did you create it?" he asks, sniffing around for a defining set he can tinker with.

• •

The stage is set now for numbers to motivate geometry. The pair of points are the universe's first real estate units. How pleasing to see your first- and second-born housed in them! You want more units, to house your entire brood. Your universe is too disorganized right now, your numbers too unmoored. What better way to introduce order than to assign a location to each one of them?

Of course, that means you're going to have to do a *lot* of generating. Starting with satisfying Two, who's already put in a request for a point like the one you gave Uno. Despite your willingness to expand your list of extra building ingredients yet again, you aren't able to conjure up another copy. Clearly there are rules you don't know yet to generating points.

The rest of the numbers begin clamoring as well. It's a golden rule of parenthood—if you give gifts to any of your children, you'd better have enough for them all. But all you can do is stare at your two—and, so far, only—points. How will you ever create all the copies needed?

You're concentrating on your pair of points, willing them to somehow multiply, when it happens. More points start to materialize—first in isolation, then in profuse multitudes. They emerge as if preprogrammed, and keep coming until they've formed an unbroken link between the original pair.

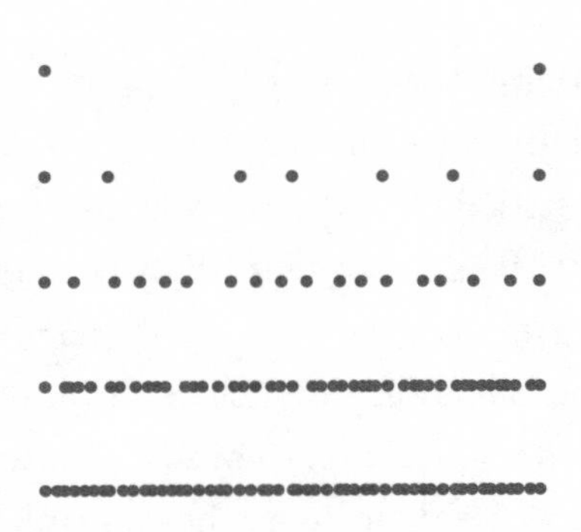

There's no "Big Bang" drama accompanying this—the birth of this first *line segment* is characterized by orderliness. Just when you think the process is complete, more points begin appearing beyond the ends to create ever-growing extensions. They proceed to generate a *line* that is endless.

You're wonderstruck by this new construct. The line brings a riveting elegance to the universe. (Let's face it, defining sets weren't much to look at.) While individual points were difficult to see, the line is more visible, lancing through the nothingness. More importantly, it contains a profusion of points to house your numbers—like an outsize condo development.

Finding a home for Two

The line has saved you a lot of painstaking point-by-point generation. But now that you have so many choices, you find yourself paralyzed. Which unit to allocate to Two? A point between those for Zero and Uno, or one outside that segment?

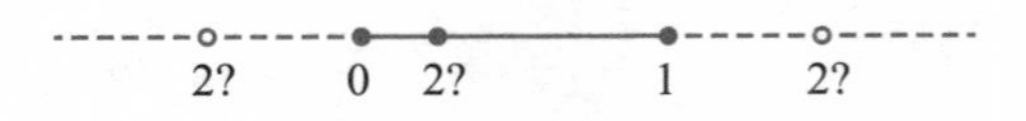

You need a systematic method of allotment, or your line will have numbers jumbled every which way. That would defeat the whole purpose of bringing order to your universe. What to use as an organizing principle?

The "greater than/less than" criterion you developed for naturals some time ago comes to your rescue. You can put 2 on the same side of 1 as 1 is positioned relative to 0, to express the fact that 2 is greater than 1 (in the same way 1 is greater than 0).* But this also presents a problem. A great multitude of points on the line lie beyond 1. Which one to pick?

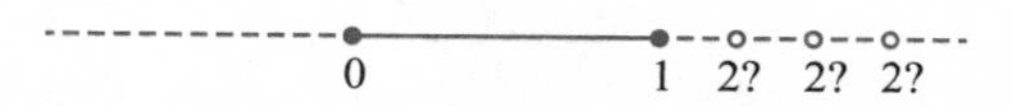

The answer will seem obvious, but scrub your mind clean of all the rulers and tape measures and *x* axes you've seen, and it might not. Remember that all you have to go by in this thought experiment is the line, the first you've ever encountered. Replay the images of points extruding outward from its ends. What is the innate attribute the line is trying to express? An attribute that increases as the line grows? Your thoughts slowly coalesce around the notion of *length*.

Yes, that's the new concept the line has introduced—one that didn't exist in your universe before. But the line's done more—it's also pointed the way to *quantify* this concept. Notice how initially only a segment got filled—the one you bookended with Zero and Uno. The difference between these two numbers is 1. What if you

---

* The point for 1 has been drawn to the right of that for 0, but this could have been reversed. Remember, there is no ambient space as yet in relationship to which anything can be oriented.

ascribe this value to the segment as the numerical measure of its "length"?

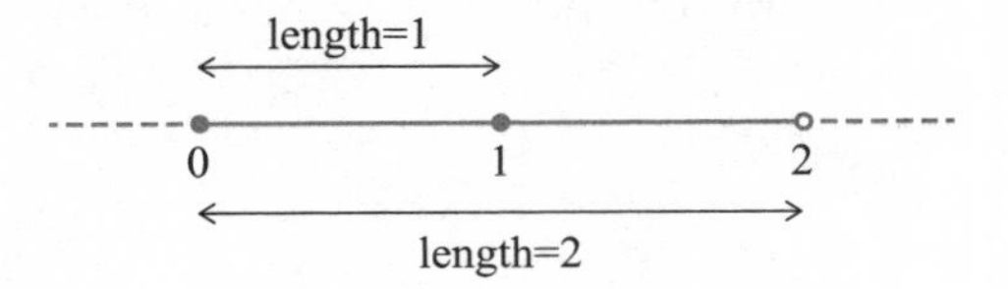

You can now imagine a copy of the original segment positioned adjacent to itself along the line. This doubled segment still has Zero at one endpoint. If you assign Two to the segment's other end, its "length" is once again the difference of the numbers at its endpoints. In other words, you've found which point to give Two!

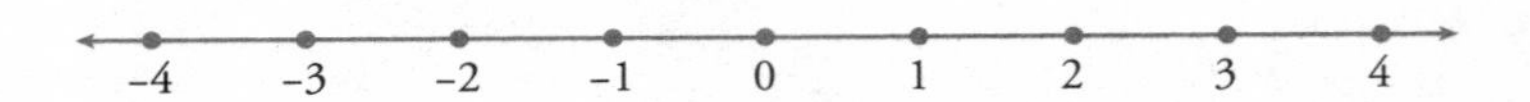

After this, the other naturals become easy to assign by marking off more segments. Another segment for Three, one more for Four, and so on—how fortunate that the line is endless! For the negative integers, measure out the same lengths, but on the other side of Zero instead. Fractions and irrationals can be assigned their own points in the interiors of segments, using the greater-than/less-than ordering again (recall that the discovery of decimal expansions made it straightforward to extend this ordering game).

SOLD OUT

Reality Realty
Condos for the real lifestyle

We do not discriminate on the basis of
negative or positive sign,
whole or fractional value,
or rationality/irrationality.

A surprise awaits you: the line has just as many points as there are reals! Every rational has its own condo, as does every irrational. Moreover, not a single unoccupied unit remains. It's as if the line was a development expressly designed for the reals. This is what I meant by saying that while geometry can't be *created* from your numbers alone, it's *inspired* by them. Think of this the next time you draw a line segment joining 0 to 1. Every number in between is being represented by your pencil tip!

Let's note another quality of this line, the fact that it's so smooth and continuous. It harbors no holes or breaks—the distinct points it is composed of have all fused together. The reals love the way it organizes them. So far, they've always floated in nothingness, with no sense of how they were connected. Now they're discovering neighborhoods of numbers they're close to, the ones from whom they're separated by tiny decimals. Taking a cue from the line, they start viewing themselves as all part of one unbroken continuum. Notice that if even one rational or irrational were missing, they couldn't have sustained such an unbroken image of themselves.

Length and distance

The original two points provide a convenient reference measure of length for your universe (much like a centimeter, say, or an inch). Recall that we set the length of the line segment between them to be 1: let's call this one *unit*. We can now describe all other lengths as multiples of this reference amount. So the length of the line segment between 1 and 4 is 3 units, as is that of the segment between −4 and −1 (length will never be negative).

With length defined, we can now introduce another essential concept often taken for granted. The *distance* between two points on the line is just the length of the line segment between them. So, the distance between −4 and −1 is 3, the same as that between −1 and −4. Like length, distance cannot be negative.*

---

* Length and distance are two expressions of the same concept. We talk about the *length* of an object, but the *distance* between two points.

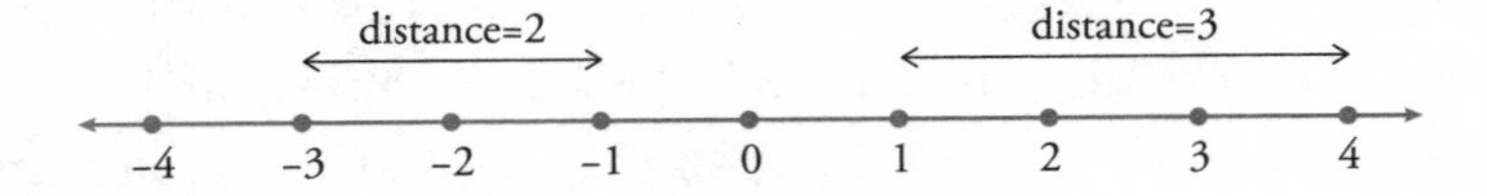

Distance allows us to give meaning to terms like "near" and "far" that we've been using rather cavalierly so far. It fills more detail into our synesthetic picture—we can now imagine our numbers strolling along the line, counting each unit as they traverse it.

Our debt to Euclid

Day 2's been fun so far, what with points and lines magically appearing out of nowhere. But to proceed, we need to examine all this drama through a more mathematical lens. What assumptions lie behind the line's emergence? What mathematical rules have been expressed? If we want to continue playing the geometry game, we have to clarify the rules, so that we can use them again in the future.

We already explained the first step, of two different points appearing in your universe. These points were "extra building ingredients," whose existence had to be assumed (pure *creatio ex nihilo* being impossible). Expect to have to add to this ingredient list each time you want to build something more in your universe.

As far as the line goes, its emergence was a manifestation of the following pair of rules, which will hold in your universe:

- Given any two distinct points, a *line segment* can be constructed between them.
- Given a line segment, a *line* can be created by extending the segment indefinitely at its ends.

If these look familiar, it's because they were formulated over two millennia ago, by Euclid himself.

Euclid realized that although these rules seem self-evidently true, one can't *prove* either of them. Given an infinite number of points, one can't very well go around establishing that a line segment can be drawn to join any pair of them. Similarly, nobody has the ability to verify that a line can be extended indefinitely. (To where? Infinity? What is that, exactly?) That's why Euclid stated these rules as assumptions—completely plausible, and very persuasive—but assumptions, all the same. The mathematical term for such an assumption is *axiom*.*

Euclid's great idea was to formulate a set of such axioms that was as small as possible, and from which he could deduce all the theorems of geometry. To do this properly, one has to list not only the basic building blocks but also any obvious-sounding relationships between them (for instance, Euclid included an explicit statement saying "The ends of a line are points"). Such a rigorous axiomatic treatment would quickly bog down our narrative with technicalities, so we're not going to follow it here. Rather, going forward, we'll assume that the following rule (which combines Euclid's first two axioms) will always hold in your universe:

✳ **Two-Point Axiom:** Given any two distinct points, we can construct an endless line that passes through them.

For brevity, we will drop the word "distinct" from now on—whenever we say "two points" we will always mean "two distinct points." Invoking this axiom in the future for different pairs of points will eventually get us all the points we need to compose empty space.

---

* Euclid divided his assumptions into "axioms" (or "common notions") and "postulates" but we will just use the term "axioms." Recall that we first introduced this term in a footnote in Chapter 2, where we noted that asserting the empty set exists could be an axiom. Another example of an axiom is the assumption that you can construct all the naturals by taking an infinite chain of successors. This is what essentially fueled our Big Bang of Numbers. "So axioms are just little math miracles?" the pope asks, and I can't decide whether he's being snarky. See the endnotes for my response.

There are two important ways our interpretation of this rule is different from Euclid's. First, Euclid took lines, planes, and 3-D space to be fundamental givens, and used his axioms only to *describe* the properties of such fundamentals, rather than *create* them. So saying a line could be "constructed" simply meant it could be drawn, by identifying points from the underlying matrix. In contrast, we actually *create* the points that lie along a line when we "construct" it. So for us, the above axiom is much more empowering.

The second difference is that Euclid stated his axioms in terms of a *straight line*, which he defined as "a line which lies evenly with the points on itself." This head-scratcher of a definition may not shed much light, but using the underlying matrix of points, he could define lines that were curved, and thus make the distinction. We, on the other hand, have built only a single line so far, so there's no way to make such a comparison. For now, therefore, we'll have to just use "line" everywhere, even though our tacit assumption is that this line is straight, and we represent it as such in all diagrams.

There's another, related aspect that's been finessed in our diagrams—the whiteness of the page or screen that shows up as background. In reality, there's no surrounding medium—you don't have the luxury of Kandinsky's blank canvas as yet. You're like an ant crawling along a wire; the lone line is the entirety of your geometric universe.

Is there a way to break out? Is there even a reason to believe that something beyond your line's confines may exist?

There is. The answer lies with the complexes.

8.

SETTLING THE COMPLEXES

REMEMBER i, THE SQUARE ROOT OF −1? THERE WAS SOME question as to why we ever defined it. Or, for that matter, complex numbers, which include every real number combined with every imaginary (like $-1 + 2i$ or $\frac{1}{2} + \frac{1}{2}i$). It didn't seem like there was any way to use such complexes for counting. What role could they ever hope to play in universe-building?

Well, the juncture has come for them to showcase their worth. Suppose complexes didn't exist, and all you had were the reals. Then the line you constructed in the last chapter would have nicely accommodated all your offspring. It's all the geometry you'd need. Your universe—which you're building based on numbers, remember—would be complete.

Complete, yes—even perfectly organized—but not particularly deep or interesting. Certainly, it wouldn't have ever reached the horizon where space gets created: you'd be forever stuck in 1-D. Physicists would smile indulgently at your baby attempt. Even the pope might allow himself a tiny instant of satisfaction (but then would feel guilty).

Fortunately, we did create i and the complexes, even though it was for a completely aesthetic reason: ensuring that the negatives have square roots just like the positives. Now we see that this abstract endeavor (which some might have regarded as a pointless flight of fancy) has a practical consequence. The complexes also need to be located, organized. They're numbers too, like the reals—so they should have their own points. "In fact, we demand it," *i* declares.

Could you squeeze *i* and her mob into the line? Even with every point on the line already subscribed by the reals, the question isn't as absurd as it seems. As we'll see on Day 6 when we explore infinity, a condo development with an infinite number of units can have surprising flexibility.

But even if such accommodation in the line were to prove feasible, it would be a terrible idea. The reason is that the reals have finally been organized in an orderly way—adding the complexes would completely disrupt things. Right now, if you traverse the line in the 0-to-1 direction, then each real number you encounter is strictly greater than all those that came before—analogously to names in a list that has been alphabetized. The complexes cannot be merged in an obvious way that preserves this increasing progression. They don't subscribe to the same kind of greater-than/less-than ordering as the reals. For instance, $3 + 4i$ and $4 + 3i$ are obviously unequal, but one can't say which number is greater or smaller. Even if you could squeeze $3 + 4i$ into the line, between which two reals would you insert it? More basically, where would you put *i* itself?

The only conclusion to be drawn is that the geometry of your universe is not yet complete. You're going to have to create something more than just the line, in order to accommodate the complexes.

It's a plane!

The first order of business is surely to settle *i*. She's always been all elbows and knees—even more so since the reals have been housed. Of late, she's taken to threatening that if nothing's done, she might have the complexes storm the line and occupy the units assigned to reals.

The only way to placate *i* is to create a new point just for her, one that isn't already occupied by a real. This means you'll have to expand your list of building ingredients again, through the following additional assumption:

* There exists an *extra point* that does not lie on the line you've created.

Barely have you had this thought than such a new point pops up. As required, it's not on the line occupied by the reals.

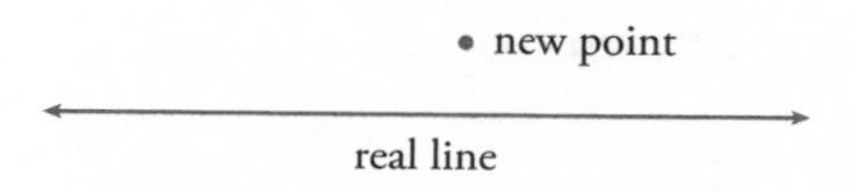

Unfortunately, all your attempts to usher *i* to this new unit fail. If anything, she gets even more ornery. "Do you think you can buy me off? You need to house *all* nonreals, not just me."

How to produce all these extra points? This is where the two-point axiom (page 79) with which you created the original line comes in. Suppose you put the new point to use as one of the two points required by the axiom. Choose the other point to be any point on the original line. Then, by this axiom, you should be able to produce a new line through this pair of points.

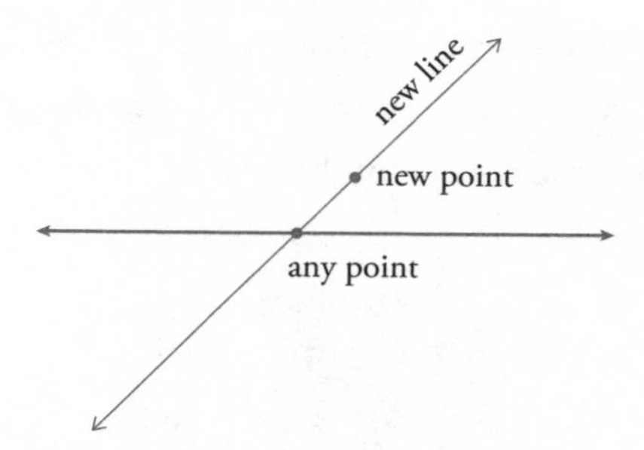

Indeed, as soon as you concentrate on these two locations, the intervening points start materializing. They fill in the segment, then extend beyond to form an endless line.

Numbers drop whatever they're doing to gawk at the emerging construct. The configuration of the two lines is spectacular, reaching far into your universe, promising further new realms to come. You're struck by how different the second line is from the first, not in composition but—you struggle to find the vocabulary—in *orientation*, *direction*. This appears to be an entirely new attribute of lines, one completely different from length.

The just-born line has given you a huge selection of new points. You realize this enables you to produce a multitude of additional new lines, using the same two-point axiom. Just select any point on the new line and join it to any point on the original line. Keep doing this again and again, for different pairs.

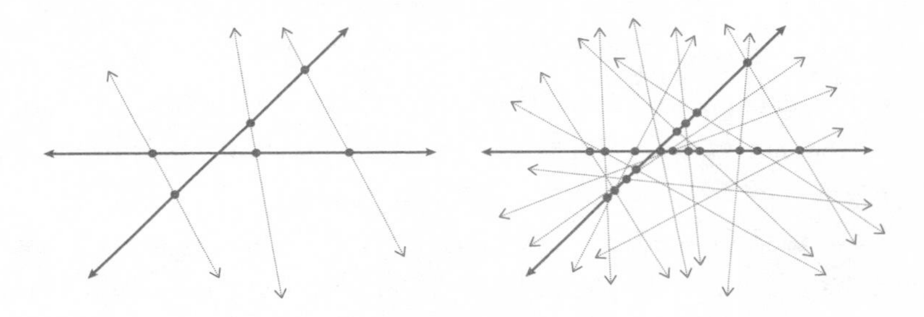

You throw yourself into this construction. There are *lots* of such pairs, so it's going to take a while. As you slog away, you wonder if you could skip some lines. After all, so many points are traversed multiple times, which adds nothing, since a point, once generated, remains the same. Could there be a shorter process that avoids so much duplication? The answer is yes, as we'll see ahead (check out the endnotes as well). For now, you file this thought away.

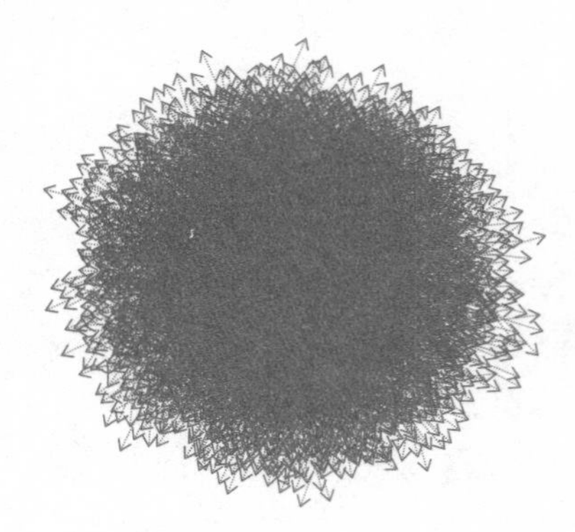

Finally, you're finished. Your efforts have resulted in a real estate bonanza—lines and lines' worth of new condo units! But there's more. Just as the individual points between 0 and 1 fused to give something more than their aggregate (a line), your newly produced lines, all crisscrossed together to create a fabric of points, also seem to have formed an entirely new construct. Something smooth and seamless, stretching endlessly in all directions, in which the individual components merge to surrender their independent identities. You call it a *plane.*

The real numbers are intrigued by this new playground. They emerge from their points to start strolling across it. How nice to no longer be constrained by the line's unidirectional trail! They love the fact that the plane has no slits or holes or residual ridges from the individual lines joining up. Every point needed is there. Zero begins to speak in poetic terms about how the universe has been transformed by this expansive new domain.

You're excited about this plane as well. But will it be large enough to settle the complexes?

A line for the imaginaries

Your youngest *i* is so pleased with what you've constructed that she names it (rather presumptuously, you feel) the "complex plane." It's all you can do to stop her from swooping down with her hordes of numbers to haphazardly occupy the new condo units. "We need to maintain order to make sure there are enough spots for everyone," you say. She grudgingly agrees to wait.

The first order of business should be to settle the imaginaries, you decide. Since every imaginary number (like *i*, 2*i*, 3*i*, . . .) is just a real number multiplied by *i*, it should be possible to accommodate them all on a single line. Could you just use a copy of the real line for this purpose? It should then be easy enough to settle the imaginaries in a row of matching, similarly spaced condo units.

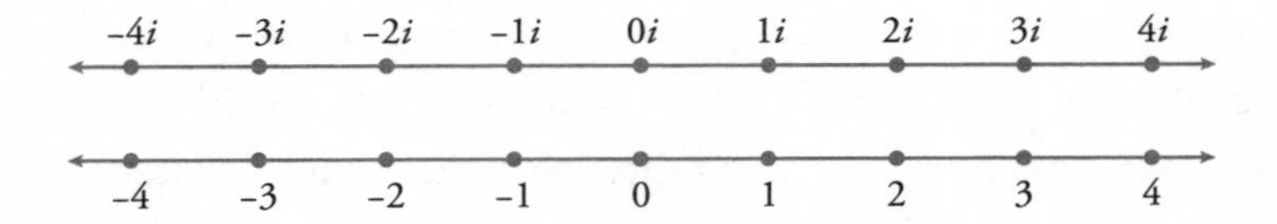

But there's a problem with your idea. The number $0i$ equals $0 \times i$, which is identical to the number 0 (which occupies the origin). You can't very well assign the same number to two different units. Especially not Zero, who's completely content being at the origin, and has no desire to occupy another point.

How to proceed? It takes some thought to figure out the key: the condo units for $0i$ and 0 have to be the same! This means that although the imaginaries can be assigned their own line, this line has to intersect the real line at the origin, the point where Zero lives.

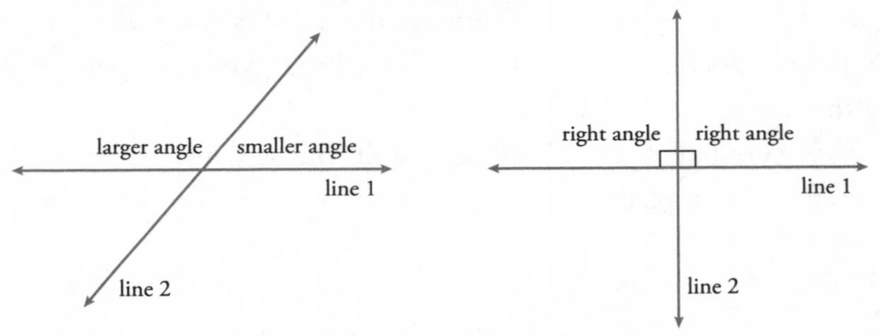

Now, you've been noticing how all the new lines on your plane have been intersecting at different "angles," a term you've coined to characterize the difference in their directions. Usually, there's a larger angle on one side of the point of intersection and a smaller one on the other. Sometimes, though, these two are equal—in such cases, you term both "right" angles, and call the two lines "perpendicular." Also, you've discovered that some lines on the plane don't meet no matter how much you extend them. You call such lines "parallel."

Let me emphasize this definition, since it will be important later. Two lines are *parallel* if (and only if) they're on the same plane and never meet.

As far as the intersection of the lines for the reals and imaginaries is concerned, making the two lines perpendicular to each other (rather than tilted at some other angle) seems to be the way to go. This will emphasize the independence of the imaginaries from the reals, something i has always trumpeted.

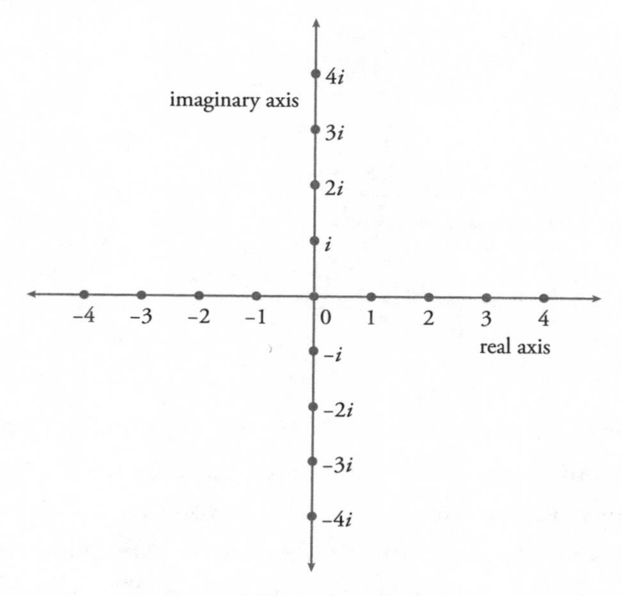

So you select the line through the origin that is at right angles to the original line. You now have an axis housing all the imaginaries, just as you have one housing all the reals. This choice elicits a nod of approval from i. But she reminds you again about all the waiting complexes.

It's a plane, revisited!

To settle the entire set of complexes, let's start with figuring out where to put complexes like $1 + i$, $2 + i$, $3 + i$, and so on, that is, the ones given by

$$\text{(some real number)} + i.$$

If the real number is 0, we just get i, to whom we've already allotted a spot on the imaginary axis. Suppose we could make a copy of the real axis and draw it through this point for i. Then it's easy to see that every complex of the type mentioned could be nicely settled on this line: not just $1 + i$ or $2 + i$, but also $\frac{1}{2} + i$, $-1 + i$, $\pi + i$, and so on.

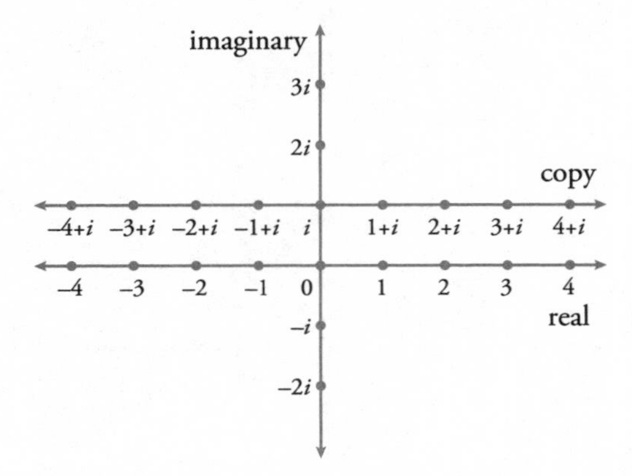

So, the question then becomes: Is it possible to make such a copy? In other words, a line parallel to the real axis, that passes through i?

You might say "obviously" yes. In fact, you may add that only *one* such line can exist, so that this parallel line is *unique*. However, notice we're talking about lines that stretch to infinity. We can't really *prove* what happens there, just as we can't *prove* that two (or twenty!) such parallel lines couldn't exist. So despite how certain we are that this can be done, we really have to label our belief an assumption. In other words, we need to have the following *axiom* handy so that we can put it to use to obtain the copy we want:

* **Parallel Axiom:** Given a line and a point not on it, we can construct a unique line that passes through the given point and is parallel to the given line.

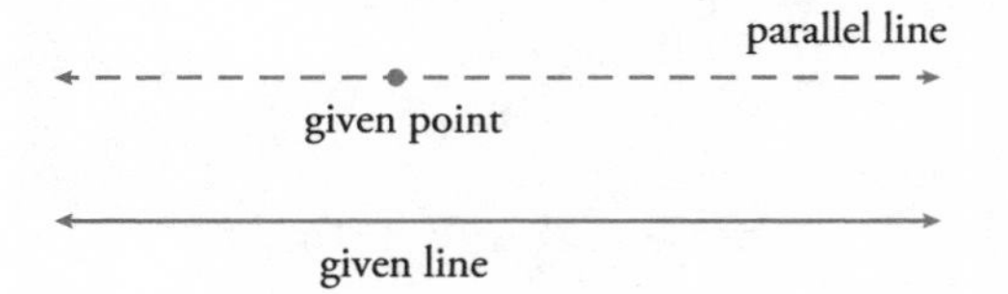

Take a good look at this axiom. Step outside our thought experiment for a moment, to convince yourself it's "true" as far as the physical universe we actually live in is concerned. Hold on to this belief, because it's going to be challenged before this Day is done.

Whether or not this axiom holds in our physical universe, we can certainly impose it on our thought experiment's universe. Then, in addition to creating the above line, we can start settling the remaining complexes. Those of the form $1 + 2i$, $2 + 2i$, $3 + 2i$, or, more generally,

$$(\text{real number}) + 2i$$

will be settled on the parallel line passing through $2i$. Those of the general form

$$(\text{real number}) + \tfrac{1}{2}i$$

will be settled on the parallel line through $\frac{1}{2}i$. And so on. By the end of this process, we've drawn a parallel copy of the real line through each and every point along the imaginary axis. Stacked up like an infinite number of matchsticks, these lines merge together to cover the entire plane.

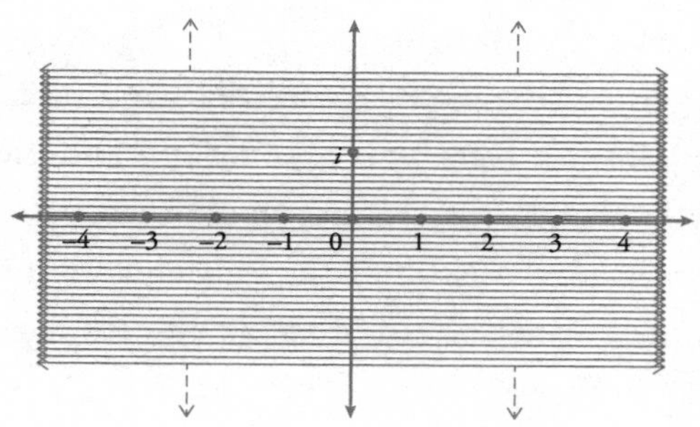

With all these parallel lines marked out, we can finally usher all the complexes onto the plane. It's easy to get them to their condo units—all they need to do is go to the parallel line corresponding to their imaginary part, and then proceed the number of units corresponding to their real part along this line. For instance, the complex $3 + 2i$ finds the parallel line through $2i$, and then proceeds 3 units along this line.

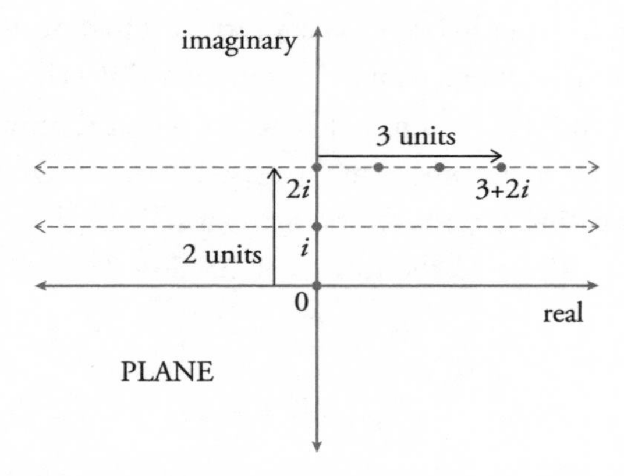

The numbers are thrilled to be settled. As in the case of the line and the reals, there turn out to be exactly as many points on the plane as there are complexes—no unoccupied condo units remain. When you see *i*, she finally has a satisfied smile on her face.

The parallel axiom brings up an interesting question. We used it to figure out how to settle the complexes on the plane. But could we have used this axiom to actually *create* the plane in the first place? The "stacking up" of parallel lines gives a tidier, more intuitive way of looking at how the plane is composed. Notice how many of the points are no longer duplicated, in contrast to all the repetitions seen when we use the two-point axiom.

There turns out to be a hitch. Recall that we called two lines parallel if (and only if) *they were on the same plane* and did not meet. So when the parallel axiom asserts that you can construct a line parallel to another, it tacitly assumes the existence of a plane containing them. In other words, to use it to *create* such a plane presents a chicken-and-egg problem.*

Despite this, going forward, let's indeed think of the plane as being generated by the stack of lines parallel to the real axis. This understanding will be helpful when we construct 3-D space.

---

* Notice also that we'd still need the imaginary axis in order to construct a parallel copy of the real axis through each point of it. To generate this imaginary axis, we'd need to use the two-point axiom.

9.

FUN AND GAMES ON THE PLANE

THE BIRTH OF THE PLANE IS CAUSE FOR GREAT CELebration. Not just for the previously neglected complexes, which are homeless no longer, but also for your universe, which can now host an array of geometric creations. Numbers flash their points on and off in jubilation.

This ability to turn points "on" or "off" is an essential ingredient in our thought experiment. We don't have materials like paper and pencil to draw anything, so this is the only way we can indicate geometric constructions. Imagine the numbers can light up their points on cue, transforming the entire plane into a giant computer screen, an infinite liquid crystal display. Geometry is about to burgeon—you can sense it as surely as you might the arrival of spring.

The universe's first shapes

The first thing you realize is how easy it is to manifest the two-point axiom. Call out any pair of different complex numbers and an entire line lights up—the one that passes through the corresponding two points. You're no longer creating the intermediate points, just activating them, so the appearance of this line is instantaneous.

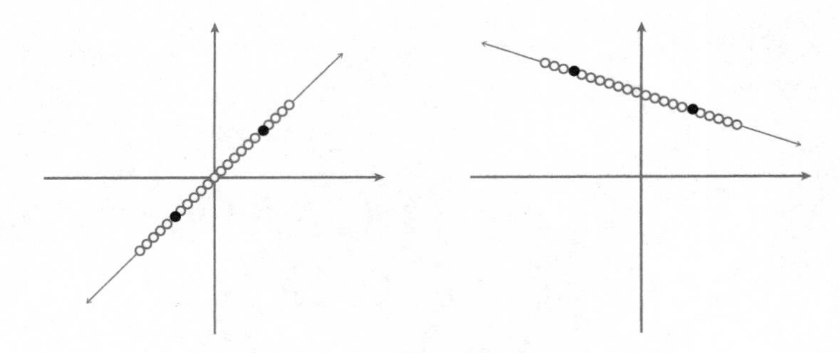

Something very interesting happens if, instead of a pair, you specify *three* different points (or, equivalently, three complex numbers). You now get three separate line segments, one from each pair. Nestling between these three segments is something quite precious—the universe's first shape! It's like an elementary life form emerging at the start of an evolutionary chain. You call it a *triangle*.*

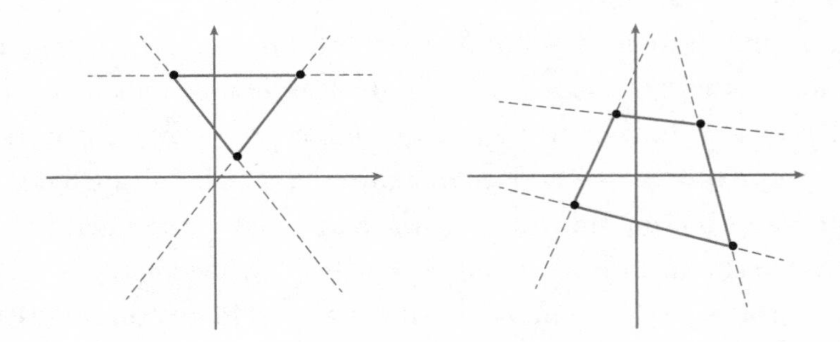

You draw quite a few of these triangles before you start experimenting with four points instead of three. Now you start seeing shapes with four sides, which you call *quadrilaterals*. These give way to *pentagons* and *hexagons* and then, as you keep taking more points, to arbitrary-sided *polygons*. The shapes start getting pretty wild—like strange, exuberant Christmas ornaments.

---

* Such triangles surely must have formed earlier, when the plane was being constructed from intersecting lines. But you could be forgiven if, amid all the hustle-bustle, you didn't notice them.

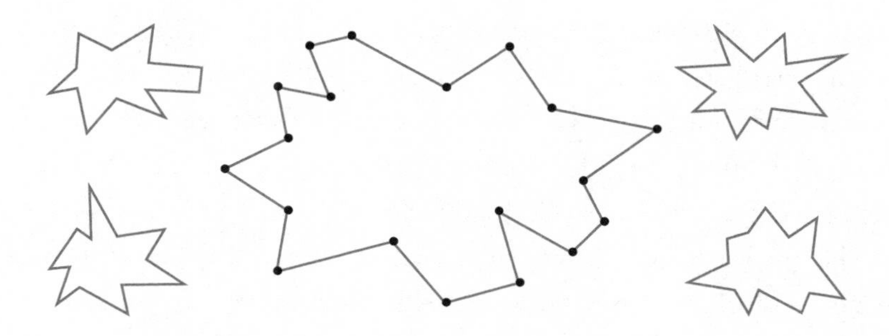

These additions change your universe. No longer is it a barren stretch of nothingness interrupted at best by isolated points and lines; now it has all these striking figures in its landscape. Each divides the plane into an "inside" and an "outside"—a new pair of concepts. You exult in the knowledge that you're doing what you set out to do—generating all this geometry using numbers and nothing else! Matching them to points has blurred distinctions; numbers transform freely into shapes, and shapes into numbers. You can even obtain a filled-in version of any figure by specifying that all numbers "inside" light up their points. This leads you to think of another new concept, called *area*. In time, you'll devise methods to calculate area, and use it to compare the sizes of different figures.

For a while, you pick numbers at random, getting a kick out of the unpredictable outlines they generate. Eventually, you decide you want to learn how to hone these outlines. Which numbers should you select to produce a desired shape? Such matching is difficult to do, and it's going to take new tools to master it. Right now, you realize you're not even sure what to aim for, which shapes would be more aesthetic.

Length and distance again

Actually, that's not true. You *do* know exactly what shapes would be more pleasing, you just can't introduce them into our thought experiment yet because they need further math. Shapes like equilateral triangles, rectangles and squares, polygons with equal sides, and the one you crave the most: the circle. These seem much more basic than all those odd-looking Christmas ornaments. Shouldn't they have been the ones obtained first?

The answer is no. Such regular shapes are actually *more* advanced in our evolutionary tree because rather than just being generated by random lines intersecting, an extra mathematical ingredient goes into their definition: length. Or, equivalently, distance.

Now, we've encountered length and distance before, but only along the original line. To proceed with our thought experiment, we need to extend these concepts to the plane. Specifically, given two arbitrary points on the plane, we have to define something called "distance" between them.

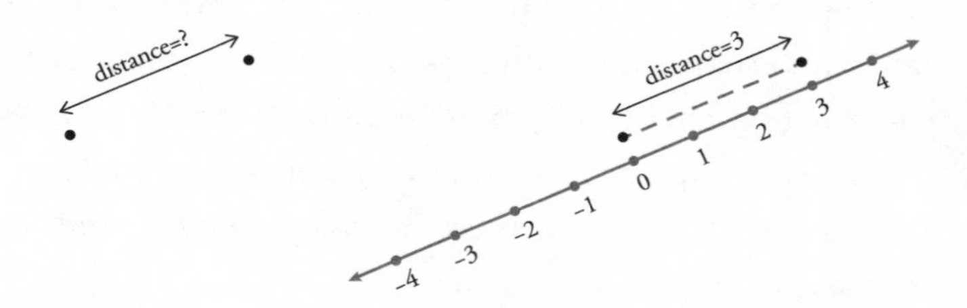

This isn't hard. Recall, there's always a line segment that connects the two points. Draw it, then measure the length of this segment in terms of your reference unit. (You could even imagine your original real axis positioned along this line segment like a ruler.) The distance between the two points is then just the length of the line segment (in units). Even the numbers can measure it, if they walk along the line and count their steps.

Did this strike you as a bit too technical? The kind of detail only a mathematician could appreciate? Couldn't I have just taken a more intuitive approach to distance instead of actually *defining* it? Not really. We'll soon see that there are several alternative formulations of distance that can be just as practically justifiable if imposed on your universe—so you have to pick one by defining it. A surprise twist waiting in your future, to demonstrate how intuition can lead you astray.

For now, let's look at the possibilities opened up with length and distance in our tool kit. Perhaps the crowning achievement is being able to draw a *circle*. The recipe is simple: light up every point on the plane whose number is exactly one unit away in distance from the origin 0. You hear gasps and applause as the universe's first circle takes shape.

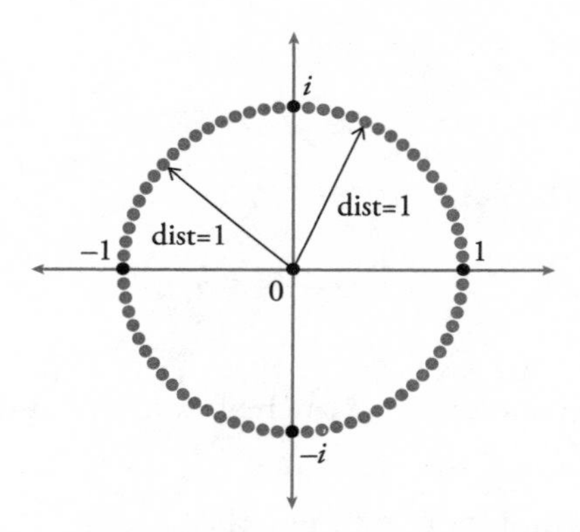

Even you are amazed that this bare-bones instruction could create a figure so radically different from the lines and polygons you've seen so far. How to articulate this difference? The current vocabulary of your universe is insufficient, so you come up with new words for the purpose. *Straight* to capture the single-minded unidirectionality of the line, *curved* to indicate the circle's repudiation of straightness. In fact, this circle seems determined to constantly keep changing direction.

The numbers are delighted to have this alternative path to stroll along. So far, they've rarely strayed from lines, which they now start calling "straight" lines.* Experiencing the circle's unwavering curl is like being on an amusement ride. They soon find the definition works just as well with different points taken as *center* and different values for the distance (or *radius*). Circular paths start getting traced out all over the plane.

* Finally on the same page as Euclid!

Meanwhile, you figure out how to use the circle to create some "more pleasing" polygonal shapes. For example, you're now allowed to introduce equilateral triangles in our experiment. But these can be tricky to draw correctly, since it's hard to pinpoint the vertices (or corner points) so that all three sides are exactly equal. With the circle, the task becomes easy. Take the line segment from 0 to 1, say, and draw two half circles of radius 1, with centers at 0 and 1. Then the point at which these segments intersect will be the third vertex of the triangle you want.

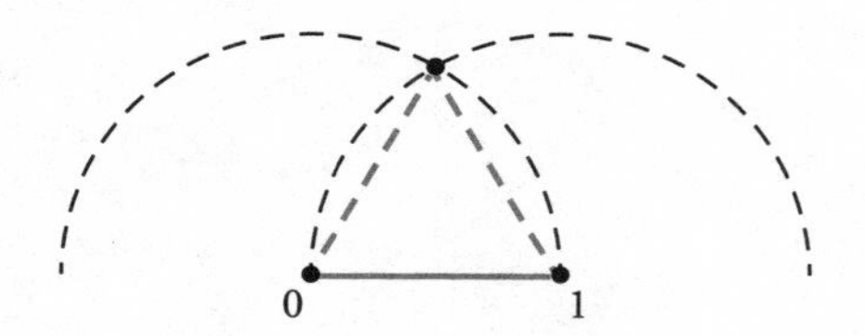

You discover other such recipes to create squares and rectangles, even equal-sided pentagons and hexagons (all of which might ordinarily involve tricky manipulations to get the sides and angles just right). We won't get into these constructions here, but they all use only straight lines and circles.

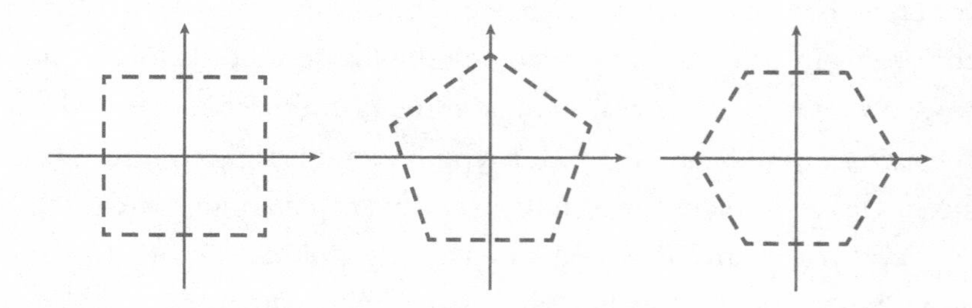

In short order, your universe is populated by a teeming menagerie of shapes, both regular and not.

10.

TWO QUESTIONS ABOUT THE PLANE

When we began this Day, we mentioned there might be alternative geometries out there waiting to be discovered through our thought experiment. It's time to take a step toward such discovery. These geometries could play a fundamental role not just in any universe you create but also in the one we live in.

What does it mean to be straight?

The above question about "straightness" arises in your mind from watching your numbers explore the plane. Ever since they realized the freedom it affords, they've been embarking on new trails, zigzagging around like skaters on ice to create all sorts of patterns. They like taking a different route each time, even if it means several extra steps just to get from 0 to 1.

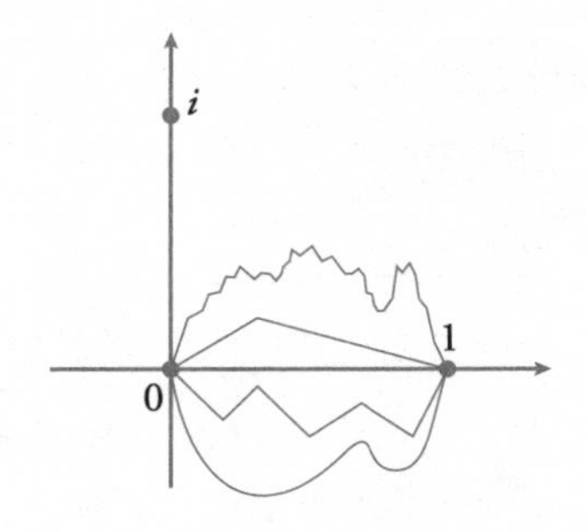

You think back to the juncture when points first welled up to link these two initial locations and form the first line segment. Couldn't the points have followed any of these "ice-skating" routes? A line with a corner, a meandering zigzag, a curved path, or perhaps some track you haven't seen yet? Straightness was an unknown concept—you didn't have any precedents to characterize it, any way to define it. And yet, you're pretty sure the prototype line came out straight. What led to this?

Could it have something to do with the intrinsic property that sets a straight line apart—the fact that it gives the shortest path between any two points on the plane? Deviate even a little, as all these numbers on strolls have demonstrated, and you end up with a longer route. Could this principle have somehow been operational, leading to your first line coming out straight, without any kinks?

Not really—for the simple reason that the concept of distance only got defined *after* the line segment had materialized to connect the two points. In fact, multiple paths connecting the two points didn't exist at the beginning, so it would be meaningless to ask which one was "shortest." Such possibilities emerged only once a plane containing the two points had been constructed, once the intervening points defining alternate paths actually existed.

Perhaps, then, it had to do with the two-point axiom you used. Perhaps it should have explicitly stated that one can construct an endless *straight* line that passes through two points. After all, this is what Euclid specified. Except, as already noted, this wouldn't have made sense, since unlike him, you were starting from scratch. You had no way back then of differentiating between "straight" and otherwise. This became apparent only once you'd developed the plane.

What might have happened if the original line segment between the first two points had emerged curved—the arc of a circle, say? You'd be none the wiser—like an ant on a wire, you had nothing to compare it with. You'd still have merrily called it a line—perhaps even a "straight line"—and that would have become the default template.

Now, if this circular line segment somehow ended up generating the same flat plane as before, then you'd be able to tell there was a shorter straight-line path between the two points it connected. You'd know then that your original line couldn't have been straight.

But what if starting with the arc generated something different—some unknown geometric alternative to the plane? An alternative such that for two points on it, the shortest distance was precisely the arc of a circle? Then you might keep thinking you'd started with a straight line. You might even label this alternative construct a "plane" just like before.

Of course, you have no idea what this alternative might be, or whether it can even exist. You try doodling with circular arcs as well as with other curves but aren't able to generate anything that fits the bill. The plane confines you, imposing a constraint you can't escape: whatever you construct is always part of the plane. This riddle is key to being able to break free of the geometry we're accustomed to. But we'll be able to tackle it only once we've constructed 3-D space.

An alternative to the complexes

Another, more straightforward question occurs to you, perhaps from memories of your math education. Why not use pairs of reals, rather than the complexes, to describe the plane? After all, this is what everyone learns in high school.

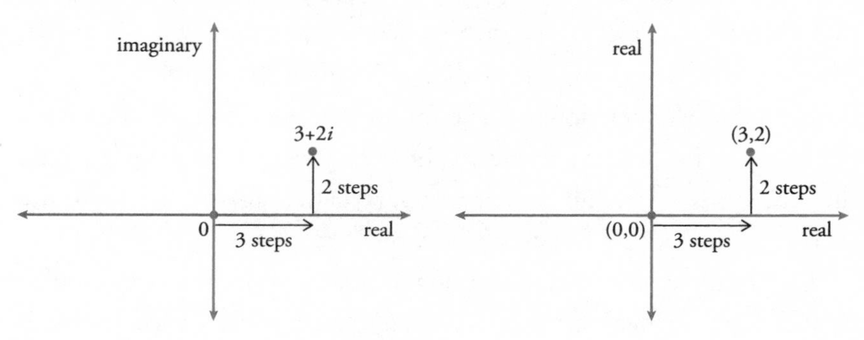

For instance, $3 + 2i$ can be replaced by the pair (3,2). This is called an *ordered pair* because the *order* of the entries matters—for instance, (3,2) is not the same as (2,3). The instructions this pair contains to reach the point in question remain the same: take three steps along the first axis and then two parallel to the second (or vice versa, as we did before). In essence, the plane now has both axes real instead of one real and the other imaginary. Figures can be lit up just the same, only, with the new numbering of the points, there's no more need to invoke the complexes. (Since these ordered pairs are coordinated with the axes, we'll call them *coordinates*.)

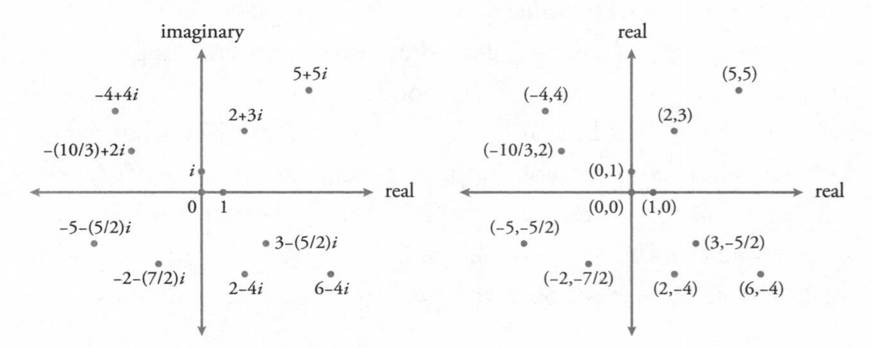

Naturally, *i* is not too thrilled about this possible change. She reminds you that without her, you might still be stuck in 1-D, might never have ventured beyond the line. It was only the effort to settle her complexes that led you to construct the plane. Is this the thanks she gets?

The reals counter that most of the plane is controlled by nonreals. Reals have been restricted to a single line—the real axis—so they have almost no role to play in geometry. The situation is so unfair, they say, that you should immediately perform the switch.

But *i* doesn't back down. She reminds you how long the complexes had to wait for housing while all those reals were living the high life in their units. "After all the exclusionary tactics against us? Let those real supremacists get a taste of the same."

Eventually, you decide not to make the change. Since every point is individual, it feels fitting to associate each with a *single* number (i.e., a complex). You feel this will make it easier to perform mathematical manipulations on the plane (a hunch that will prove correct on Day 4). Besides, if you evict *i* and her complexes, where will you resettle them?

Of course, *i* couldn't be more delighted to keep ruling the roost over her "complex plane."

11.

SPACE OR BUST

THE PLANE WITH ALL ITS SHAPES IS WELL AND GOOD, but we still haven't created space. That's what this Day's main goal is supposed to be. It's fine to ponder conundrums, but we need to pick up the pace. Remember, the universe's components emerged in a fraction of a second after the big bang. Even Genesis took just six working days.

The third dimension

A philosophical question stands in our way. The straight line accommodated the reals, the plane took care of the complexes. With all our numbers settled, what possible motivation could there be to create space? Is there some other, larger set of numbers we haven't constructed yet that needs to be housed?

The answer is no—the complexes don't have a natural extension that would occupy space.* So such a justification fails. Does this mean we're destined to be forever stuck in the "Flatland" we've built?

Far from it. We have an irresistible motive to drive us forward. Curiosity. The pioneering spirit. The first two points gave us the line.

---

* Mathematicians have come up with more esoteric extensions like the so-called quaternions (which we won't get into here) but not one that compellingly motivates the construction of 3-D space.

A point outside the line gave us the plane. What will happen if we conjure up a point outside the plane? The train's been set in motion, we need to see what the next stop is.

So let's have you add one more extra building ingredient to your list by making the following assumption:

* There exists an *extra point* that does not lie on the plane you've created.

It takes but an instant for this new point to materialize, distinct from the complex plane.

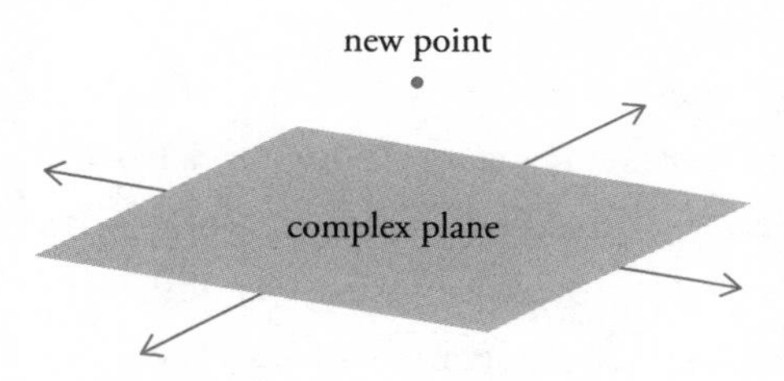

Now, relying only on the two-point axiom, we can follow a procedure similar to the one we used for the 2-D plane, to construct 3-D space. But since the method of stacking up parallel lines like matchsticks was so much more intuitive, let's use that for our demonstration instead. For this, we'll need to additionally assume the parallel axiom, whose analog for planes is: given a plane and a point not on it, we can construct a unique plane that passes through the given point and is parallel to the given plane.

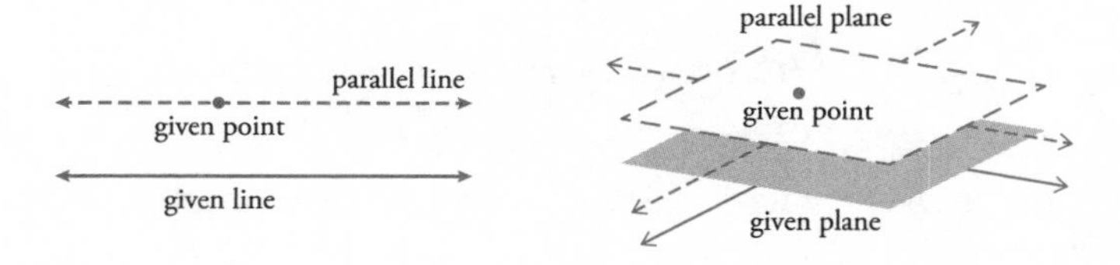

With this rule in place, we're set to go (see the endnotes for explanatory details). Start with a line through the extra point perpendicular to the complex plane. Then, through each point on this line, you can create a plane parallel to the one containing the complexes. These planes are as endless and easily generated as lines were earlier. They stack up neatly, not like matchsticks but like a book's pages.

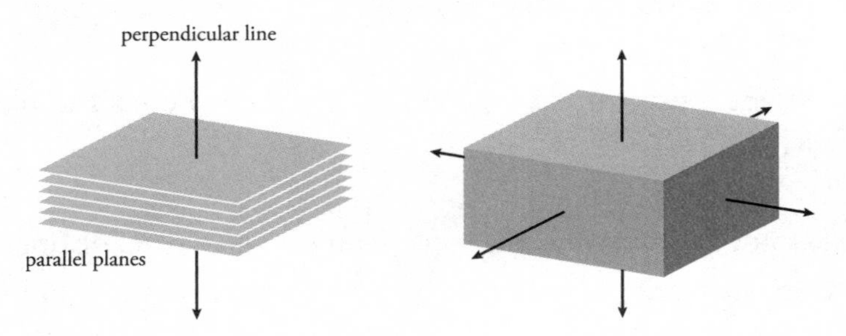

Run through every point on the perpendicular, and a wondrous transformation occurs. The planes merge with one another seamlessly to create an entirely different realm. Whereas before, you could experience two dimensions, there is now a third as well (call it height). The new construct seems to fill everything, be everywhere, inviting you and the numbers to surrender yourselves to its boundless extent.

You've finally reached the promised land of *space*. It banishes the last stultifying notions of nothingness, giving your universe a medium in which to envision itself. Zero, especially, is delighted. "Before, I used to despair about being so alone in the emptiness. Now when I feel the darkness coming, I imagine gathering space around me and comforting myself."

The exile of the complexes

Space, being everywhere, has encompassed the complex plane. Axes and all, with Zero still at the origin. This plane is the only part of space that's inhabited, with complex numbers occupying their points, turning them on and off. The rest of your new real estate is bare, undifferentiated.

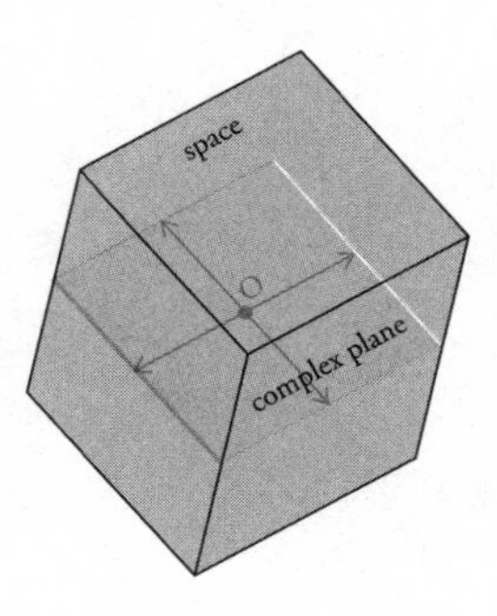

You realize something needs to be done about all the extra condo units you've generated. Without more numbers to fill each empty position, there's no way to even reference individual points in space, much less light them up to create new shapes. The problem is there *are* no new numbers waiting to be created.

Although i immediately tries to colonize these new spots, none of her strategies take. The reals, already resentful at how they've been overshadowed on the plane, are alarmed by her new annexation efforts. They approach you with an enhanced version of their earlier plan about assigning a pair of reals to each point on the complex plane. Except they'd now add a third number—a zero—to the list. So $3 + 2i$ wouldn't just become (3,2) but (3,2,0), and the origin would now be (0,0,0). The third component would measure how far above (or below) the complex plane the point lay, measured along a new axis perpendicular to the previous two.

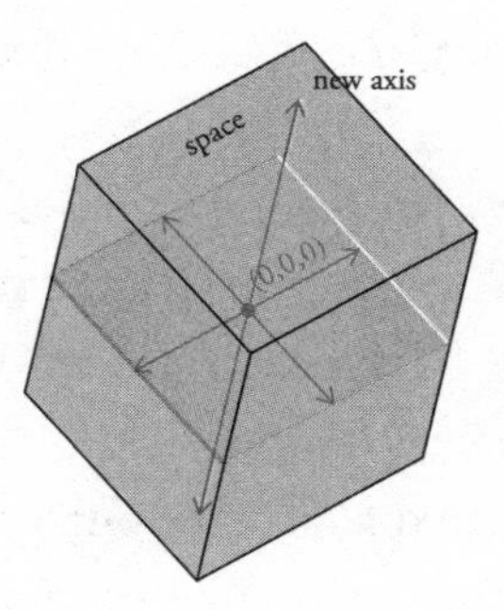

For example, to match the triple (2,3,4) to its point, you would take two steps from (0,0,0) along the first axis, then three steps parallel to the second, and finally, four steps parallel to the third. You can similarly match a triple to each point in space, so that each point would now have three coordinates.

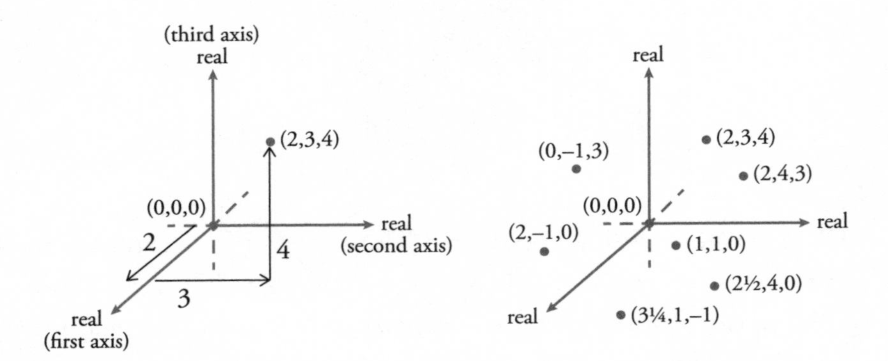

The proposal is brilliant. You have little choice but to go with it, especially since *i*'s attempts to expand beyond the plane have all failed. Ordered triples of reals are all you need to populate your new real estate!

Unfortunately, this means such triples will need to be instated at each point of the complex plane as well—which, after all, has now become a part of space. Otherwise, the old numbering with complex numbers won't be consistent with the new numbering with triples used everywhere else. It falls on you to do the dirty work of evicting the complexes.

There is disbelief in *i*'s stare when you explain that even though you've constructed so many new planes, they're all taken by the newly formed triples. Which means the complexes have nowhere to go. "So we're to be homeless again," *i* exclaims. "After all we did to create the plane!"

The actual eviction is wrenching. Several complexes break down and have to be assisted out. Others have to be ripped from their condo units. "I suppose we'll have to learn to be a tribe of wanderers again," *i* says. She leads her complexes away.

After that, you see *i* frequently in your ruminations, her features windswept and hardened, spurring her emaciated tribe across impossible terrains. This is your burden of guilt to bear, you tell yourself—the collateral cost of discovering space. Who knew your synesthesia would lead to such nightmares?

12.

AN ALTERNATIVE GEOMETRY

ONCE YOU HAVE SPACE, A NEW GOLDEN AGE OF geometry dawns. The first inkling you get is when you ask all points the same distance away from the origin to light up. It turns out that there are now many additional such points compared to 2-D. So instead of the routine circle you're expecting, a surprising new shape appears—a sphere. It's as if the circle has replicated itself many times over to form a 3-D incarnation of itself. The numbers all marvel at this sphere's perfect roundness, its flawless symmetry.

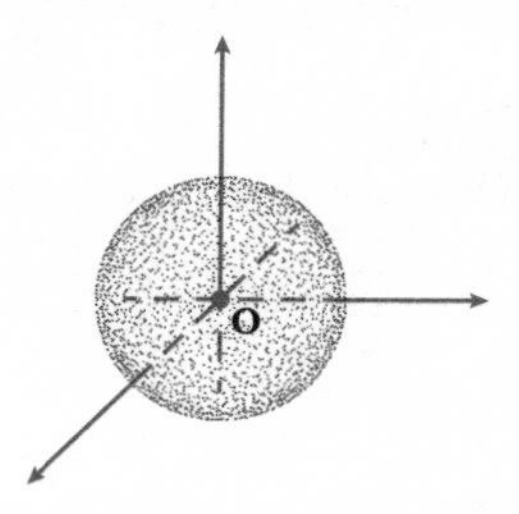

You create many new shapes—perhaps the most interesting of which are formed by stacking together planar figures (just as you stacked entire planes to create all of space). For instance, filled-in circles (discs), when stacked, lead to a cylinder, rectangles to a brick, and squares to a cube. You come up with a new concept called *volume* to compare their sizes.

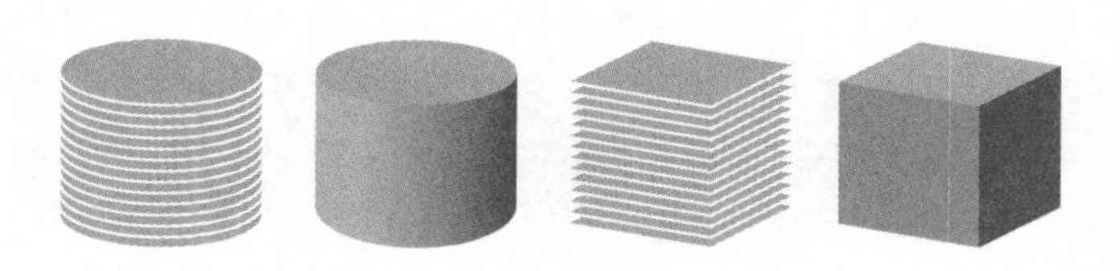

The number triples get into the construction boom as well, lighting up points on their own. Usually, this just results in shapeless forms, but every once in a while, they chance upon a tetrahedron, a pyramid, a prism. You watch them play hide-and-seek in the crevices of strange crinkly figures that nobody knows what to call.

Slowly but surely a constellation of shapes lights up—the heavenly bodies of your cosmos.

What it means to be straight, revisited

You're starting to get rather comfortable with space. There's a pleasing depth to the figures being created, a shapeliness that wasn't quite possible in 2-D. You're confident your 3-D will satisfy both popes and physicists. How nice to relax in this new realm, take a well-deserved break before resuming the creation of your universe.

But a previous question keeps nagging you. It's the issue of the line segment that materialized to join the first two points. What if it had not been straight? What if it had been the arc of a circle, for instance? Could this have given a changed geometry that was still viable? One in which the shortest distance lay along circular arcs instead of straight lines?

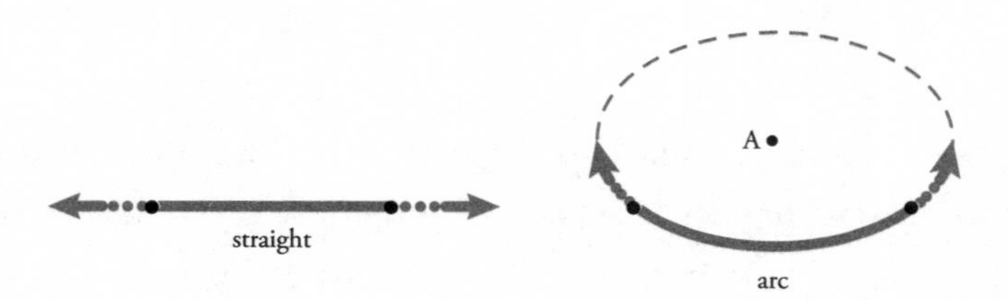

The only way to be free from this question is to tackle it. So imagine an arc connecting the first two points, and picture this segment extending. As points bubble forth from the ends, they don't form an infinite line, but curl around to meet in a circle instead. For definiteness, picture this circle as having radius 1 and center at some point A. This will be your new "line" (or "straight line"). You now want to figure out what kind of "plane" will result if you start with such "lines."

Here's an easy way to get to the answer. Notice that the plane is simply the aggregate of all possible lines that can be drawn on it. This means that if you define "line" to be "any straight line that lies on the plane," and then draw each and every such line, you'll generate all the points that constitute the plane (in fact, each point would be duplicated many times). Of course, this description is circular and self-referential, so does not serve as a practical recipe for construction. But were you able to follow it, it would readily yield the plane.

Now suppose we redefine "line" to be "any circle with center A and radius 1." Then, in analogy with the above, if you draw every possible such circle, you should generate the new version of the "plane."

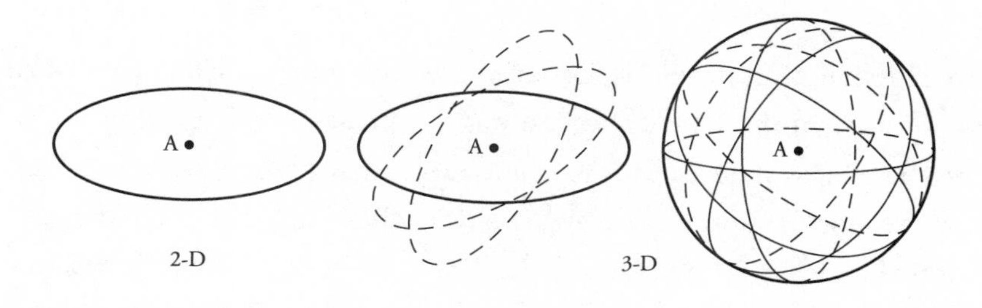

If you confine yourself to 2-D, this condition is too restrictive since there is just one such "line"—the original circle. However, when you get to 3-D, many such "lines" can be drawn, each a circle with center A and radius 1. Draw all such "lines," and it's not too hard to see what you get: the unit sphere with center at A. This, then, must be the counterpart of the plane!

In other words, you've answered the question that's been gnawing at you. Could you get a geometry different from the plane if "line" stood not for "straight line" but something else? The answer is yes. Change the definition to a circle and you'll generate the surface of a sphere instead! In this spherical geometry, "lines" will now be "great circles," that is, circles with maximal radius 1.

What about the two-point axiom (about a "line" existing to join any two points) and the construction based on that? Do these carry over for the sphere? Indeed, they do. A version of the axiom will still hold, and you can generate the sphere using the same step-by-step procedure you used in Chapter 8 for the plane (details in the endnotes).

The parallel axiom, on the other hand, fails on the sphere. Note that any two great circles will always meet—at two diametrically opposite "poles" of the sphere. Consequently, with this definition of "line," we see that "parallel lines" on a sphere don't even exist! So, given a "line" on the sphere, and a point not on it, you can draw *no* parallel "line" through the point!

What we really mean by a line

There remains an unresolved question. Is the shortest distance between two points on the sphere always along the great circle (i.e., "line") joining them? It is indeed. This is why long-haul airplane routes try to follow the globe's approximate great circles, which can take them over the Arctic Circle, even the North Pole. The mathematical term to describe such lines along which shortest paths lie is "geodesic." Thus, on a plane, the geodesics are straight lines. On a sphere, they are great circles.

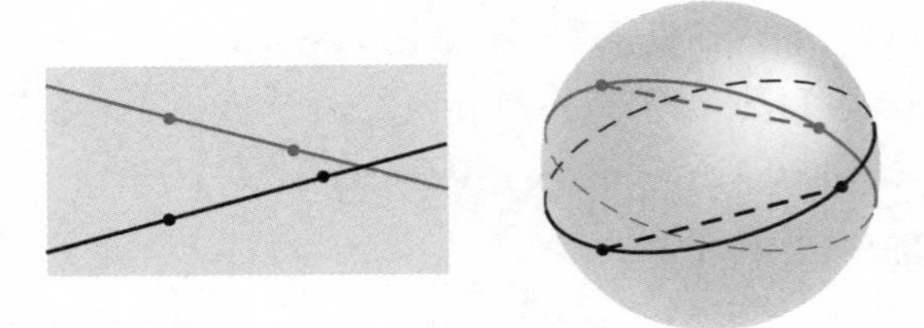

Now, you might object to this. After all, you could burrow through the sphere and have a shorter straight-line path right there. But this thinking is fallacious. You can't go straight through because the surface is all you have, all you've constructed. You're like an ant living on the sphere's surface, or an airplane navigating the globe. For you, the internal points don't exist.

Notice, therefore, that the very definition of distance between two points can change: it depends on whether you consider them to lie in 3-D space or consider them only in the context of the sphere's surface. In other words, "distance" is not something to take for granted—it depends on geometric context. For instance, it's 11,000 kilometers between New York and Beijing along the Earth's surface, but only 9,700 kilometers if you bore a hole to tunnel through. Clearly, we're not so different from ants when it comes to interpreting our planet.

The above considerations help us finally pinpoint what, exactly, we mean by "line/straight line," a term we've purposely been somewhat loose about so far. Quite simply, we mean "geodesic," that is, a path in the geometry (2-D plane, sphere, 3-D space) along which the shortest distance lies. So our two-point axiom, substituting "geodesic" for "line," should really read:

* **Two-Point Axiom:** Given any two points, we can construct an endless geodesic that passes through them.

There are two qualifications to this. First, "endless" could mean "infinite" as in the case of straight lines on a plane, or "finite but unending" as in the case of great circles on a sphere. In the latter case, the geodesics are endless in the sense that they wrap around to form a continuous, unending loop.

Second, there has to be an underlying geometric structure (for instance, a 2-D plane, a sphere, or 3-D space) in which the two points lie, and to which the geodesic corresponds. That's because geodesics, as we have seen, don't exist in an absolute sense but depend on the geometry (specifically, on how distance is defined). As explained above, if we just consider the surface of the sphere, they're great circles. If we change our geometry to be 3-D space, then the geodesic between the same two points becomes a straight line going through the sphere.*

All this brings us to a bigger question. We've seen that the plane and the sphere are two possibilities of geometries we might construct. Are there others?

The answer is yes. There's a radically different family of geometries waiting out there to be discovered. This family emerges when you use a different interpretation of straight line/geodesic to connect the first two points. These new geodesics are neither the planar straight lines that generate a plane nor the great circles that generate a sphere. Proceeding as before, if we drew every possible geodesic of this type, this would result in our new geometry being generated. Moreover, in this new geometry, the two-point axiom would once more be satisfied: given any two points, an endless geodesic connecting them could be constructed (and in addition, it would be unique).†

All this sounds good, so why haven't I told you what these new geodesics are? The problem is they're rather complicated to define. Worse, if I were to try to draw them, they wouldn't correspond to any curve (like a straight line or circle) that you'd easily recognize.

Fortunately, there's a much easier approach to understand this new geometry. One that cuts through all the abstractions of Day 2 and gives us something tangible we can hold on to. I mean this quite literally. We're going to take a break from the mental gymnastics I've been putting you through and construct this new geometry through crocheting!

---

* There's a third qualification. Given two points, is the geodesic between them unique, or can you draw more than one through them? For the case of a 2-D plane (or 3-D space), there's only one straight line between points. Euclid didn't state this explicitly, but that is, indeed, what he intended. For the case of a sphere, the geodesic will be unique in all cases *except* when the two points are at diametrically opposite points of the sphere. Then, like the lines of longitude on a globe, an infinite number of lines can be drawn through the pair. If one excludes such diametrically opposite points from consideration, then the geodesic between any two points will again be unique.

† So employing this axiom could be an alternative way to get the geometry, using the same technique of connecting points that we used to generate the plane (and sphere).

13.

CROCHETING YOUR UNIVERSE

Crocheting can be thought of as a real-life approximation of our Day 2 activities. A crocheted fabric is made up of individual stitches the same way a plane is made up of individual points (stitches being grossly larger, of course). Indeed, the first crocheting lessons one learns are how to make a chain stitch (the analog of a line) and how to create a patch from rows of stitches (analogous to how a plane might be constructed out of stacked lines). With enough yarn for an infinite number of stitches, one could hypothetically crochet an entire plane.

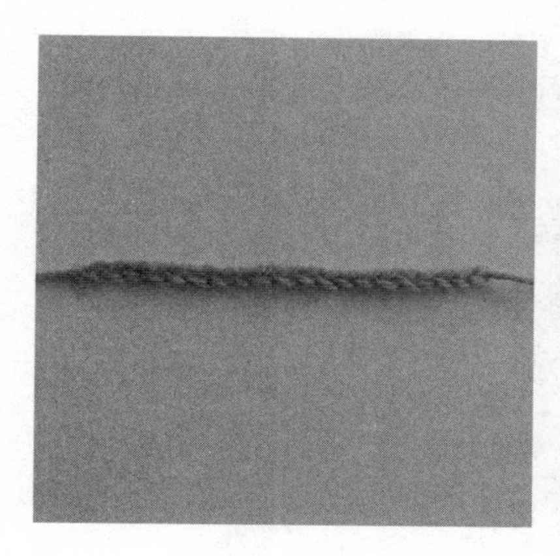

Creating an infinite plane via a rectangular pattern is problematic. That's because for a rectangular patch, one crochets along a row to the end and then works backward for the next row. Such a zigzag pattern becomes unworkable if each row is to be infinite. It's much easier to picture starting with a circular patch, constructed by increasing the number of stitches in each concentric ring by the same fixed amount. For instance, a typical recipe might start with an innermost ring of 6 stitches, followed by succeeding rings of 12, 18, 24, and 30 stitches. Continue such rings all the way to infinity, and you get the plane!

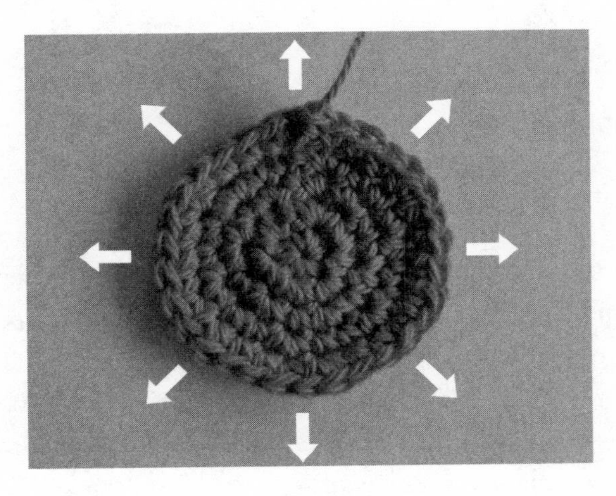

Crocheting websites often point to a problem that can make the resulting disc not quite flat. *Cupping* occurs when you don't add enough stitches as you progress outward. The way to fix this problem is simple: you just need to keep adding a few extra stitches as you go along.

What happens, though, if you purposely keep skimping on the extra stitches needed in each ring? Your cupping will get very noticeable, because you're not supplying enough yarn to form a flat plane. Suppose you first slow down the required increase, and then actually start *decreasing* the stitches in each ring. You'd then be able to approximate something very familiar: a sphere! For example, this is what emerges with 10 rings of size 6, 12, 16, 20, 22, 22, 20, 16, 12, and 6 stitches.

This suggests a new way to interpret the spherical geometry we created in the previous section. We end up with it when we don't supply material (i.e., points!) at a fast-enough rate to create a "flat" geometry, that is, a plane. Notice that the crocheted sphere uses only a finite amount of yarn, as opposed to the endless supply that would be needed for an infinite plane. Due to this paucity of material, everything gets drawn together. This causes lines that might have been parallel to converge at polar points. Also, the sphere's surface bulges outward, with the result that triangles drawn on it have extra area, and their angles sum to more than two right angles. (On a plane this sum would be exactly two right angles.)

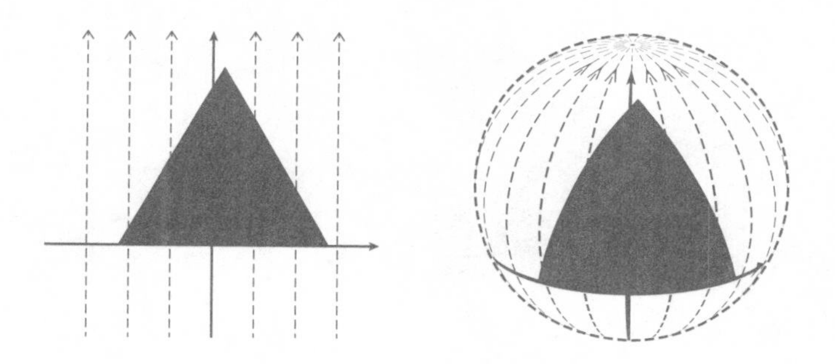

A new geometry

There's a second danger that crocheting websites warn about. *Ruffling* occurs when there are *too many* stitches. Now there's more material being added than can be incorporated into a flat plane, which creates a ripple effect at the edges. Rather than pulling together as for a sphere, the patch tries to open up, expand, so that it can deal with all the extra yarn you're putting in. In addition to the ripples, it curves into a characteristic saddle-like shape.

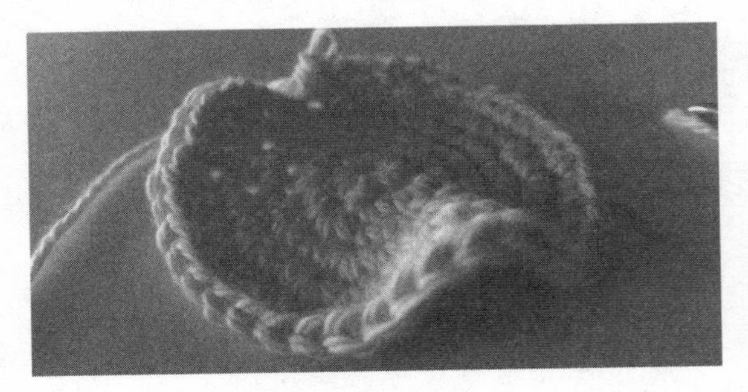

This brings us to the new family of geometries I promised. Suppose you were dealing with points rather than stitches, and adding them at a faster rate than your usual flat plane could handle. Then the resulting geometry would behave *oppositely* to a sphere—opening up and expanding, rather than pulling together or closing up. Parallel lines would now diverge due to this ubiquitous expansion, getting farther from each other as they extended across the surface. Triangles would now constrict inward, leading to their areas shrinking, and the sum of their angles being less than two right angles.

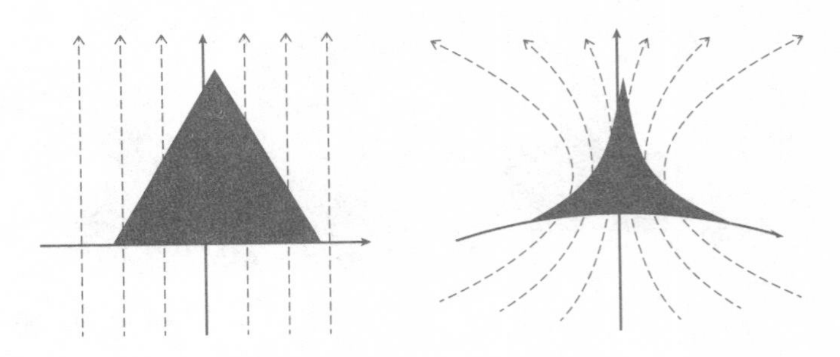

The quintessential example of this geometry is the so-called *hyperbolic plane*. This may have remained an abstract shape existing only in the minds of mathematicians had it not been for Professor Daina Taimina of Cornell University, who showed how to bring it to tangible life through crocheting.* Her constructions, such as the one pictured below, depend on increasing the number of stitches in each ring at precise exponential rates. The resulting surface now curves into much more extravagant ruffles, and also displays the characteristic saddle shape when you examine any patch. (Note that crocheting gives only an approximate representation; the actual hyperbolic plane is infinite.)

If the above shape reminds you of corals and sponges, that's no accident. These animals depend on filtering the water around them for sustenance, so they need to maximize the surface area of their body. Hyperbolic geometry, with the extra area that goes into its gen-

* Her work, remarkable for the way it makes abstract math accessible, has inspired this entire chapter.

eration, is the perfect evolutionary solution for them.* (See the endnotes for more on corals and crocheting.)

Other marine animals, like sea slugs, flatworms, and nudibranchs, derive a very different benefit from hyperbolic geometry: they undulate their ruffles to propel themselves through the water. Studies have shown that their shapes are more flexible than other configurations and require less energy to produce motion.

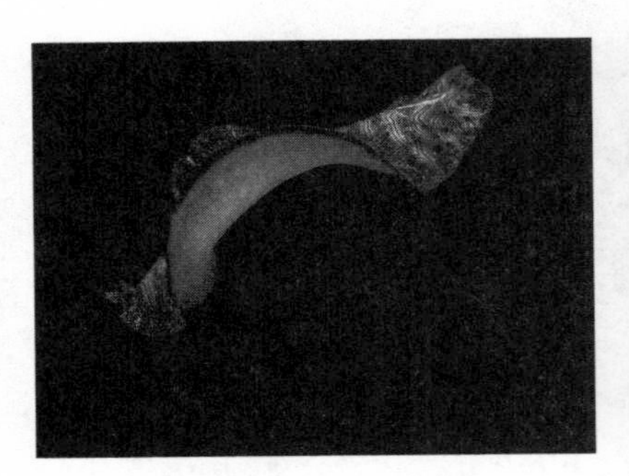

---

* In contrast, an organism may sometimes want to *minimize* surface area, since (for instance) that's what exposes it to possible breaches. In such cases, it would want to opt for the "opposite" of hyperbolic geometry, i.e., spherical geometry. This is one reason why spores and seeds are often spherical: by minimizing surface area, the biological cargo inside is protected.

There are many more instances of hyperbolic geometry in nature (e.g., the ruffled tops of mushrooms, the curly leaves of lettuces). Again, form is usually tied to function—there is often some advantage this geometry confers on the biological specimen.

Whither parallel axiom?

Although the formal definition of a straight line (i.e., geodesic) is best relegated to texts on geometry, once we have a crocheted representation of the hyperbolic plane, it's actually quite easy to see what such a line looks like. Just as folding a piece of paper symmetrically reveals a straight line on a flat plane, so also folding the crocheted piece (without stretching it) reveals a straight line on the hyperbolic plane. The pictures below illustrate some examples. Each line contains the shortest-distance path between any two points lying on it. Remember, this is the shortest distance along the hyperbolic plane, rather than in full 3-D space (the latter would just be a regular straight line).

The picture on the right reveals something else interesting: the parallel axiom fails again! Recall that it failed on the sphere because in that geometry there were *no* parallel lines. Now it fails because, given a line and a point, you can draw *more* than one parallel line through the point. (Recall that we defined "parallel" to mean lines that never meet, rather than lines that are the same distance apart everywhere.) On the photograph, we've labeled two such lines, both parallel to the given line, but you can easily mark more of them in between these two. In fact, due to the hyperbolic plane's excess area, you will have an *infinite* number of lines through the given point, all parallel to the given line.

If you're wondering why I keep bringing up the parallel axiom, it is because this is one of the most famous propositions in all of mathematics, one that tantalized mathematicians from the time Euclid introduced a related version as his fifth and final axiom. Euclid himself felt his fifth was too obviously true to be an axiom, and so should be deducible from his four preceding ones. Consequently, he held off deploying it all the way up until Proposition 29 in *The Elements*. For more than two millennia after, many others also tried and failed, like Euclid, to show that the fifth had to be a consequence of the four remaining axioms.

Only in the nineteenth century did mathematicians realize the parallel axiom could *not* be derived from the other four, the reason being that a perfectly valid alternative to Euclid's flat geometry existed in which the fifth axiom failed, even though all the others held. This discovery, of an abstract version of the hyperbolic plane, caused a revolution in the mathematical world. It contradicted the deeply held belief that mathematics existed only to describe physical reality, and that therefore anything as intuitively "obvious" as the fifth had to be a logical necessity. (The fact that hyperbolic geometry also described nature, and had been adopted by sea creatures a half billion years before humans had any inkling of it, was realized only later.) Mathematicians were now forced to accept the fact that their subject had to be built up axiomatically, and that it was possible to get a different and completely consistent mathematical universe when you changed the starting assumptions. In particular, Euclid had been correct in enshrining the fifth as an independent axiom, almost as if in anticipation of "non-Euclidean" geometries to come, where it would be violated.*

Since then, various other deeply held beliefs embraced by mathematicians have fallen, all victims to the fact that the subject has to be built up axiomatically. Which is all well and good, you might say, but surely mathematicians' abstract and highly esoteric beliefs have nothing to do with your life?

They do, as it turns out. Our reality is intimately tied in with whether the parallel axiom holds in our universe, as we shall see.

---

* Perhaps this vindication is what inspired the following lines from arguably the most famous poem on mathematics, "Euclid alone has looked on beauty bare," by Pulitzer Prize–winning poet Edna St. Vincent Millay:

> *O blinding hour, O holy, terrible day,*
> *When first the shaft into his vision shone*
> *Of light anatomized! Euclid alone*
> *Has looked on Beauty bare. . . .*

The above metaphor, comparing mathematical creation to cosmic inception, is certainly one we've been inspired by in this book (witness the Big Bang of Numbers). Euclid is deservedly cast as the messiah, the one who first "anatomized" mathematics by recognizing its elementary building blocks: axioms.

14.

THE FOURTH AND HIGHER DIMENSIONS

TO FURTHER UNDERSTAND THE SIGNIFICANCE OF the parallel axiom, let's look at how it might relate to our proverbial ant. Initially, this ant was restricted to a one-dimensional wire, but last we checked, it had been upgraded to the surface of a sphere. Let's say it's still living on a surface, but the type may have changed: it's now on either a sphere, a flat plane, or a hyperbolic plane (perhaps it wandered into the crocheting basket). How can it tell which of the three it's on?

Clearly, the parallel axiom would give the answer. If the axiom holds, then the ant's on a plane; if not, it's on either a sphere or a hyperbolic plane, depending on how the axiom fails (no parallel line, versus an infinitude of them). The problem is that it's impossible to physically verify the axiom, given that one needs to make it all the way to infinity to check that two lines on the surface are parallel. So this is not a practically feasible method.

Now, if it was you or me on the surface, we might be able to tell just by observation. For instance, a popular way suggested for schoolchildren to verify the Earth is round is to watch a ship come into view—the mast appears first, then the deck, then the hull (on a flat surface, everything would appear at once). Let's assume, however, that our ant is constrained to its surface and unable to detect anything in the perpendicular (height) dimension. Then this strategy isn't an option. Neither is taking an airplane ride to look through the window for a curved horizon.

One thing the ant could perhaps try is to measure the sum of the angles of a triangle (assuming this is a very smart ant who knows how to measure angles). On a flat plane, the sum would turn out to be exactly two right angles; on a sphere, it would be more; on a hyperbolic plane, less. While such measurements might be theoretically possible, it's safe to say that even the canniest ant would have a hard time discerning that its universe's geometry was curved.* That's because while a sphere or hyperbolic plane needs 3-D to exist, living on its surface is essentially a two-dimensional experience. This is something we all recognize as humans (and the reason "Flat Earth" societies still exist). How often do we truly become aware of the curvature of the planet we're on?

Which brings us to a fundamental question. Could there be a higher-dimensional analog of such spherical or hyperbolic curvature? Could the 3-D space of our own universe be "curved"? If so, this space might "live" in 4-D, in the same way curved surfaces like spheres and hyperbolic planes "live" in 3-D. If we led a four-dimensional existence, then we could perceive this curved space, understand it, map it—as easily as our present 3-D selves can see and handle a sphere. But being constrained to 3-D, we're like an ant in 2-D, who's unable to tell whether its surface is curved. In other words, even if our 3-D space was curved, we would still experience it as essentially flat.

Fortunately, we're not constrained by the limits of what we can experience as our physical selves. Now that the idea of curved 3-D space has been raised, we can explore it further in our thought experiment, as a possible contender for our universe-in-progress. Perhaps the first step is to construct a fourth dimension so that your 3-D space has something into which it can curve. In other words, a 4-D

---

* For a sphere, a geodesic will return to itself, so this could be a possible way of telling: if the ant walked along a "straight" line and realized it ended up returning to where it started. Make the radius large enough, though, and this becomes increasingly difficult—the ant, sadly, expires before completing its odyssey.

setting in which 3-D space has the freedom to mold or twist or curl itself into a different shape.

A four-dimensional elephant

It's an ancient Buddhist parable: a group of blind men relying only on touch to figure out an animal entirely unfamiliar to them—an elephant. One man, feeling the trunk, says it resembles a snake; another, touching the ear, thinks it is a fan; the tusk feels like a spear to a third, the leg a tree trunk to a fourth, and so on. The point of the story is the elusiveness of reality—the elephant's true physicality remains inaccessible to the men. And yet, I would hope that the aggregate of their observations gives them at least a partial understanding of what constitutes an elephant.

4-D space is *our* elephant. While we can never attain full awareness of it, we can, as we shall see, at least catch a few glimpses. After that, it's up to us to stitch together a personal image that, though necessarily incomplete, gives us enough to hold on to. This is the best we can expect for many concepts in math, which are often too divorced from everyday experience to easily visualize. Remember, there are different ways to assimilate ideas—just because you can't see something (say, a fourth dimension) doesn't mean you can't accept and appreciate its existence intellectually, by leading your mind through a sequence of logical steps. In other words, through a skill we've been developing in this book: abstraction.

With 4-D, though, we have to go one step further: not only can we not see this elephant, we actually have to give birth to it! This turns out to be easier than expected, because we already have a logical sequence, based on the parallel axiom, for its creation. We saw how this axiom could be used to get from 1-D to 2-D by laminating together parallel copies of lines, and also to get from 2-D to 3-D by layering parallel copies of the plane. To get to 4-D, use this same recipe, but instead stack together parallel copies of 3-D space. That is, use the abstract underlying pattern that's been figured out.

Accordingly, start with the following additional assumption, similar to the ones made earlier:

* There exists an extra point that does not lie in the 3-D space you've created.

Although physically impossible for us to realize, this is perfectly plausible at the conceptual level, given all our successful precedents in lower dimensions. Indeed, such a point instantly pops up in our thought experiment, distinct from all the points that constitute existing space.

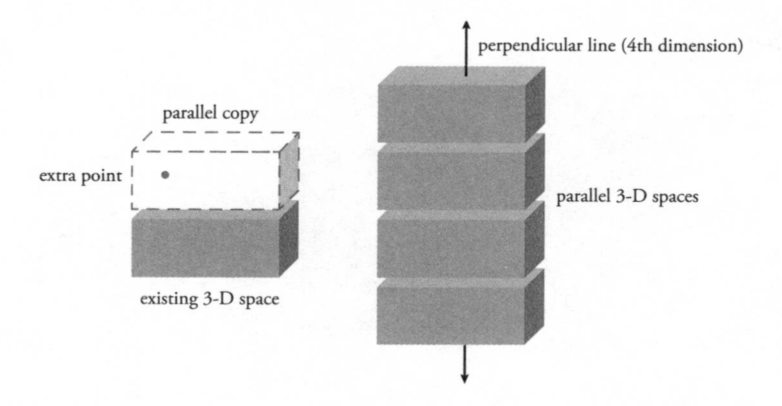

The parallel axiom now allows you to create a "parallel" copy of 3-D space through this extra point. Also, with the help of this extra point, you can obtain a line that's perpendicular to all of 3-D (in analogy with what you did in lower dimensions). In other words, there's now a fourth direction, perpendicular to the familiar three associated with 3-D space. This is the direction in which you will now start stacking a succession of parallel copies of your original 3-D space.

Since this fourth direction has to be perpendicular to everything, there's no reasonable way to visualize it (you'll notice I've shamelessly cheated in my illustrations). But the procedure works just as well as it does in lower dimensions: the copies of 3-D merge to form 4-D space. You find yourself embedded in this new 4-D realm, like a tiny decoration inlaid in a glass paperweight.

Glimpses of the beast

What can you say about your new 4-D environs? Suppose you're free to move around in them, and can look at your old 3-D space from a fourth-dimension vantage point. What strikes you at once is that you have X-ray vision. Gaze at a closed box with opaque sides containing a ball, and you can see not only all six sides of the box but also the ball inside.

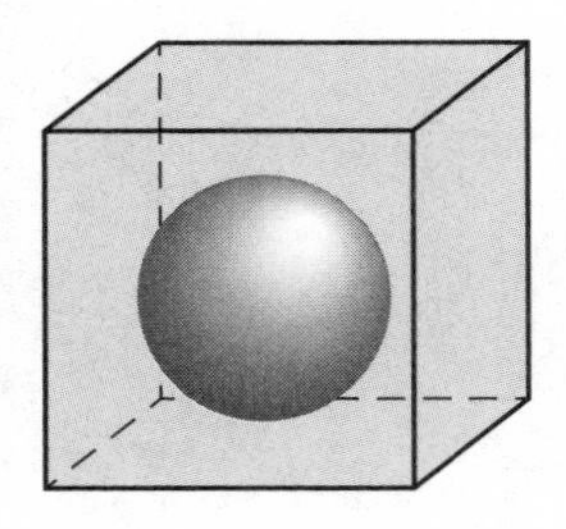

This is puzzling at first, but let me give you an analogy to explain what's happening. Say you were a 2-D being, like one of your ants, confined to a plane surface. Replace the ball in a closed box with a flat disc in a closed square. If you encountered this, you'd be able to discern just one or two edges of the square at a time. You could go all around the perimeter of the square, but since you're flat, you'd never be able to peer inside to find out what lay there. That's similar to what would happen if you encountered the 3-D box as your own 3-D self—you'd be able to see a few sides of the box, but not the ball inside.

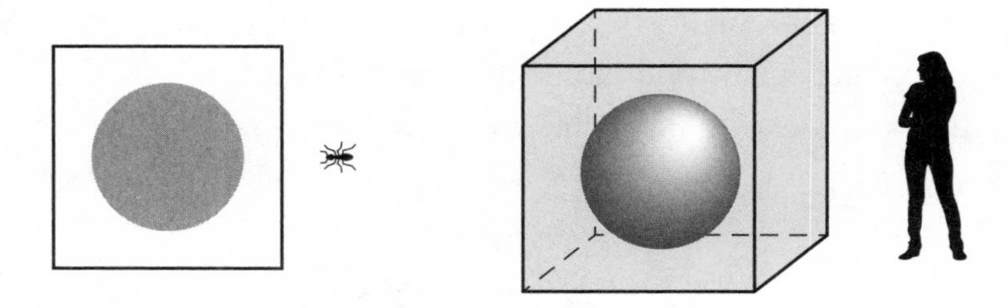

But now think of yourself as an airborne ant, somehow able to gaze at the entire plane from above. From this position, you'd be able to see all four sides of the square, and also the disc enclosed in it. The same goes for when you're a flying version of your 3-D self, able to gaze down from the fourth dimension. You can now see the box in its entirety, along with all its contents. Your 3-D universe has been splayed open like a butterflied chicken; you can view all its insides.

There's another 4-D effect. When you talk to your numbers, they have no idea where your voice is coming from. Since they're still in their original 3-D slice, you're invisible to them (as is anything outside their realm). You can move around unseen like a space alien—an effect you thought possible only in science fiction. For a while, you have fun playing God—telling the numbers in grave tones that you're displeased with their constant mischief and they'd better make amends. Finally, you reveal yourself, dropping into their world out of nowhere, it seems to them. They're unnerved for some time after that.

You realize there's an easy way to populate 4-D with numbers. Just have each point—which is now a 4-D condo—occupied by an ordered list of four numbers, rather than three.* For example, (2,3,4,6) would be assigned the point reached by starting at the origin, proceeding 2 units along the first axis, 3 units parallel to the second axis, 4 parallel to the third, and 6 parallel to the fourth. The origin would now house the quadruplet (0,0,0,0). This matching offers the easiest way to grasp

* Again, the continuation of an abstract underlying pattern learned from lower dimensions.

4-D space intellectually, the one I use myself. It's simply the set of all locations that can be pinpointed this way with four coordinates—no need to grapple with impossible mental images!

How to visualize curved space

Having 4-D under our belt allows us to clear up some issues regarding curved 3-D space. At the most basic level, since we can't really visualize 4-D, we shouldn't expect to be able to visualize curved 3-D space either. The reason is that the fourth space dimension is along an unseen direction perpendicular to the three visible to us. When 3-D space curves, it does so into this dimension, which is why we can't observe this effect.

However, the crocheting analogy is still very useful. For 3-D space, instead of more crocheted *area* being added, we can think of *volume* being piped in (imagine a squeeze bag, like one for mortar or pastry cream, being used for this). Inject just the right amount of volume, and you'll get 3-D space in the standard Euclidean flavor. This space is infinite and what we call "flat." But pipe in too much material, and the excess will manifest itself as *ruffles* that curl into the fourth dimension. There's also the possibility of a deficiency of volume, in which case the space *cups* into the fourth dimension, and can close in on itself like a sphere. This higher-dimensional version of a spherical shell will have finite volume.

The above options are by no means exhaustive—for example, space could be flat in some regions and curved in others. You could also sculpt it into doughnut-like or more exotic shapes, so that there are holes in it.

Suppose we restrict the choice in our thought experiment to just the three possibilities of flat, spherical, or hyperbolic 3-D space. Then, as in the lower-dimensional case of surfaces, the success or failure of the parallel axiom pins down the geometry we choose. We can also make the distinction based on triangles: the sum of a triangle's three angles will equal two right angles in flat 3-D space, be more in spherical 3-D space, and be less in hyperbolic 3-D space.

Could these tests work not just for our thought experiment but also to check whether the 3-D space we physically live in is curved? We've seen the parallel axiom cannot be practically verified, and so it doesn't help. But how about the idea based on measuring a triangle's angles? Both mathematicians and physicists (stretching all the way back to the legendary nineteenth-century German mathematician Gauss, it is said) have tried this, by measuring the angles of actual physical triangles formed between distant mountain peaks or even between stars in space. Other, more sophisticated experiments have also been performed to determine this. No overall curvature has ever been detected, which means that if there is any, it must be smaller than the uncertainty error of these experiments (estimated at under 0.4 percent). In other words, our universe's underlying space geometry is almost certainly flat overall,* but a very small probability of curvature remains.

The stakes are high. If our space has even a tiny bit of positive curvature everywhere (as a sphere does), then it would curve in on itself, which means it would be finite and bounded. To be infinite, it would have to be flat or curve the other (hyperbolic) way. So the answer to this curvature question, a question arising directly from the parallel axiom, may determine whether our universe is finite or could be infinite. This is an essential attribute of our existence, of our "origins and creations," as Fixmer's quote in the introduction put it. And it hinges on mathematics.

---

* Our universe consists mostly of empty space, with actual matter like stars and galaxies occupying only a tiny fraction of it. These might cause small local bumps or deformations (as we shall see on Day 5), but our question here is: Ignoring these anomalies, does our universe have any curvature?

Which geometry to choose?

Now that you've figured out how to construct all these different versions of space, a mischievous thought strikes you. The conventional choice for your universe-in-progress would be flat 3-D space, but what if you defied expectations? What if you went whole hog for a fully 4-D universe instead?

Neither God nor the physicists would likely be too happy. You can't help but smile imagining the look on their faces. Wouldn't the intellectual challenge of retooling the big bang or Genesis do them good, though? Haven't they been resting on their laurels long enough? You could sell the idea to them by touting all the extra space (a whole dimension's worth!) they'd have for their creations.

The numbers would certainly enjoy their new digs—all the fun they could have lighting up new 4-D objects! The more you think of it, the more sold you are on the idea. 3-D seems passé—a brave new 4-D universe awaits. You'll be immortalized as its discoverer, its architect. True, it might initially make stakeholders grumpy, but they'll get over it. You'll come up with a 4-D version of cricket that actually keeps you awake! The pope and you will share a good laugh over the 3-D past, as you lounge in the redesigned 4-D Vatican, over mocktails.

You're getting excited about this 4-D idea when a thought stops you. If you can create four dimensions, what prevents you from creating five? All you'd need to do is laminate together copies of 4-D space to accomplish this. You can already represent it without even building it. After all, each time you write down a set of five coordinates, say (1,2,0,1,4), you refer to a point in 5-D space.

Why not go even higher, to six or seven or even more dimensions? Constructing these might take effort, but it would be a straightforward repetition of the recipe. Is there a particular number most suited to the space a universe should exist in? You're suddenly paralyzed by all your choices. Where do you stop, once you've learned to drink from abstraction's heady cup?

What dimension is our universe?

Fortunately, the underlying goal of our thought experiment comes to the rescue. We're ultimately trying to create *our* universe—not some science fiction version of it. The world we live in (counting only space, not time) is three-dimensional, as far as anyone can tell. So that's the dimension you're obliged to choose.

Hold on, physicists will say. Deciding on our physical universe's dimension is not so simple. They'll point out that superstring theory, which models elementary particles as arising from the vibrations of tiny strings, requires space to have *nine* dimensions. An earlier version, called bosonic string theory, required *twenty-five*, and other dimensionalities have also been proposed. All this might sound very esoteric, but clearly, your idea of creating arbitrarily high dimensions was not so outlandish.

However, there's a qualification. These extra dimensions arise from the mathematics needed to make physics theories work, rather than from direct observation. There is no experimental evidence so far to prove their physical existence. Notice, also, that the choice of 3-D does not preclude the possibility of a 4-D, 9-D, or even 25-D reality out there, with our universe restricted to just a lower-dimensional facet (the way an edge or face is a lower-dimensional facet of a block). Also, the universe could still acquire more dimensions in the future. Some theories have it transitioning from 1-D to 2-D to 3-D, and currently trying to grow a fourth space dimension.

Given that physics offers so many racy speculations, let's wrest ourselves away and go for the most conservative option: 3-D space. Flat, to boot—which means with no ruffling or cupping, because just the right amount of volume has been piped in.

The pope's relieved not to have to hire a 4-D mocktail mixologist. In all his years of religious study, he's never once come across God mentioning dimensionality or curved space. Even geometry arises most notably in relation to a certain astronomer silenced by the Church—history on which the pope would rather not dwell. Flat 3-D suits him just fine—in fact, he'd venture to say God will be OK with it as well.

Physicists are less pleased. They poke around dubiously at the 3-D space you've selected, tweaking its knobs, kicking its figurative tires. True physical space will emerge only during the big bang, they warn. Whether it will flow smoothly into your mathematical mold as you hope remains to be seen, since the extra dimension of time will cause various complications.

That's OK, I reply. We'll take our chances. A universe built on pure concept gives us great flexibility. With all the possibilities we've established, the choice can always be changed if physical reality demands it.

So there we have it. We've made our selection. Our space will be three-dimensional and flat, just the way Euclid would have wanted.

Day 3

ALGEBRA

Developing a language to communicate with Nature

15.

OUTSOURCING

AS YOU TAKE STOCK OF OUR PROGRESS SO FAR, A thought begins to bother you. You've spent the first two Days working on the underpinnings of your universe. Day 1's numbers and arithmetic operations will undoubtedly help in formulating this universe's coming laws. Day 2's choice of 3-D space has set the foundation on which everything will get built. The squares, circles, spheres, and other shapes you designed will serve as templates for future physical objects. All in all, a pretty good tally for only two Days' work.

The problem is that everything so far has been entirely conceptual. That's great in terms of flexibility, but you worry whether math can ever break free from the realm of ideas. Will it ever enable you to construct anything tangible? A planet, for instance? Or even an atom?

The answer, unfortunately, seems no. Mathematics is great at designing models and blueprints, at giving detailed instructions for construction. But it can't actually produce physical bricks or mortar, or do the building itself.

Wasn't that the promise I made in the introduction, though? In fact, in the very title of this book? Have I been leading you down the garden path all along?

I haven't. Read the fine print in Chapter 1 and you'll see I gave myself an escape clause. My promise was to have you build the universe using only numbers and math—BUT with a few helpers. Indeed, we've already encountered some: the empty set we needed to create the naturals, as well as the axioms and "extra building materials" (like points) needed for geometry. Now we're going to need another external agency—a contractor. The kind you'd hire to do the actual construction, were you building a mansion. Or a planet. Or an atom.

In case you feel a contractor bends the rules too much, think of all the great historical builders. Herod the Great of Jerusalem; Constantine of Rome and Constantinople; Shah Jahan, who gave us the Taj Mahal. As a nod to a certain VIP reader, let's also include Julius II and the other popes responsible for the Vatican. Who among these luminaries ever wielded a pickax or even a chisel?

No, the reason they're credited as builders is because of their vision, their drive and resourcefulness, the fact that they made it happen. It's the same with this project. Mathematics gives the universe its design, its inspiration—makes it happen. The case for awarding building credit to math is clear-cut.

So, let's accept, without prejudice, the idea of hiring a contractor who'll handle the actual construction, much as builders have done in the past. We're on the verge of introducing matter, so we need someone fast. Fortunately, I've found someone perfect for the job: Nature.

What does the name conjure? Snow-capped mountains and verdant valleys, forests lush with flora and fauna, expanses of sea and sky and stars. Or perhaps mythical nymphs and dryads, garlanded mother goddesses, Hindu devis with multiple arms. In her many forms she encompasses all the beauty and power that mortals ascribe to God. She'll be able to literally move heaven and Earth to build you the universe you want.

But she's also a science maven. "Physics" comes from *physis*, which means "nature"—the Greek word has deep scientific and philosophical significance attached. So when we appoint Nature as contractor, we're also allowing the possibility that whatever she builds follows strict scientific rules, with no miracles involved.

This is why Nature's the ideal choice for our project—because she conveniently embodies both the alternatives readers may want. She can represent either God or science, depending on what you believe in—or even some combination of the two if you're nonbinary on the subject. (Genesis *and* the big bang—why not? Who said humans have to be consistent?)

Keep two things in mind. First, whether Nature represents religion or science won't matter, since she's only going to follow the blueprints you give her. Her output, in other words, will be controlled by the math you generate (*how* she accomplishes things is another matter). Second, she's never done this before. She's appearing only now in our thought experiment, so hasn't seen *any* mathematics, not even numbers. It's going to be up to you to train her. In particular, you'll need a language to communicate with her about what you want for the components of the universe. Developing such a language is what most of this Day will be about.

You might wonder why I couldn't match you with a more math-savvy contractor. The reason's simple—there's nobody else around. Take heart, though—with her pedigree, Nature's bound to be a fast learner.

Meet your contractor

A swarm of particles billows in. Although they're simulations, both you and the numbers can feel them as if they've already jelled. Nature's going to experiment with them as she paves the way for tangible forms. They're as annoying as gnats. You try to track their paths, but can't ascribe any order to the patterns they make. The numbers try to catch them, without success. As suddenly as they appeared, the particles disperse, to reveal Nature making her entrance.

How to picture her? Let's resist anything too religious or scientific. Perhaps a single discreet pair of angel wings, combined with a tasteful radioactive glow to her hair. You'll need to deal with her as a contractor, so you don't want to be cowed by an image too exalted or flamboyant. Think of her as someone down-to-earth and workperson-like—a mason, a carpenter.

You introduce her to the numbers. She's intrigued by them (as they are by her) but doesn't immediately see what they might be good for. "One, two, three," she repeats, to memorize their names.

It happens when some particles fly past. "One," Nature exclaims, pointing at the first. "Two, three," she calls out, gesturing at the subsequent pair. Did she just discover counting? Realize how numbers could be associated with objects? You see this insight light up her face.

The numbers realize it too. Your rabble of offspring are electrified at being connected to external objects. So far, their life has been all about play. Are they being offered a chance to do something deeper with it? Does a new world await somewhere outside their condo units?

Smitten, they start calling her Mother Nature. You realize the numbers have never acknowledged your parenthood, so Nature's sudden elevation to this status stings a bit.

Straight lines

You embark on the first step in Nature's training—how to draw a straight line. It's a construct she's never seen before, having not been around while you created geometry. Part of your motivation is simply that she'll straighten out all those chaotic paths traced by her particles, which have been driving the numbers and you crazy. You draw her attention to the plane formed by the first two axes. "Specify a pair of positions and the line joining them will automatically light up." You call out two points to demonstrate.

It doesn't work. The points turn on, but everything in between remains unlit. You prod some nearby points, and repeat your command, but nothing more gets activated. Calling out other pairs of points also fails the same way. Nature looks on, unsure of what's going on.

You have no idea, either. It's not as if the two-point axiom could have stopped being valid. Technically, though, the axiom guarantees only that a line between two points *exists*, not that it has to light up. So the axiom itself may not be to blame.

What you are encountering is a plot twist. The automatic illumination so far has been just a temporary aid to help foster geometry. To proceed with our thought experiment, this bonus has to end. You need to develop a more self-sufficient way to draw lines, not depend on some unexplained agency to facilitate it.

Fine. You'll activate the line between any two endpoints yourself. You can individually call out each and every intervening position's coordinates, so the numbers will know which points to light up. This turns out easier said than done. The process is messy and time-consuming, the segment ends up looking like the edge of a very irregular saw.

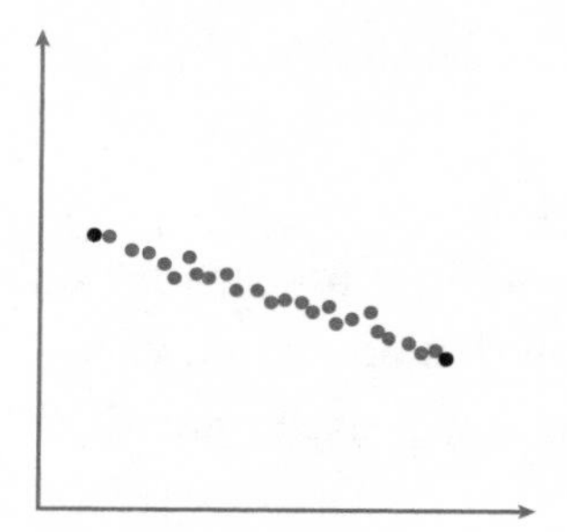

The problem is there are too many points too close together to choose from. Figuring out precisely which ones lie on the line is very difficult. You have only your visual estimation to go by, and that's not enough. Both Zero and Uno also give it a go, and the results are just as kink-filled and childish. The kind even a proud parent might think twice about displaying on a refrigerator door.

You're stumped on how to proceed. How will your contractor ever construct a physical line if you can't even convey to her what a line is?

16.

THE JOY OF x

So here's your predicament. You can no longer generate line segments by calling out their endpoints. The era of free lunches—which you didn't even realize you were benefiting from—is over.

For a long time, you stare at the blank slate of the plane. Will you ever be able to draw a straight line again? Will you be able to teach Nature? It occurs to you that no matter what, a pair of lines—the two axes—are always going to be easy to identify.

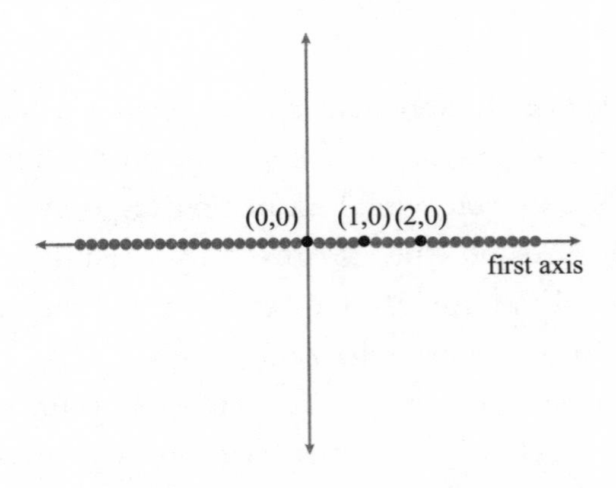

Taking a closer look at the first axis, you see something interesting: each coordinate pair* has 0 as its second entry. Which means you

* Since you've restricted attention to the plane, it's easier to just concentrate on coordinate pairs. There might still be a triple of numbers at each point on this plane, but the third entry is always 0, so ignore it.

could light up this line manually if you wanted. Not by visually gauging which points would make it straight (which didn't work when you tried it), but by activating only those points with 0 as their second number. Suppose you turned on *all* such points—not just (0,0), (1,0), and (2,0), but also (−4,0), (2.3, 0), ($\sqrt{2}$,0), and so on. Wouldn't that do it?

There's only one way to find out. "We're going to play a game," you announce to the numbers in their units. "Light up your points if your second entry is 0; otherwise, stay unlit."

For an instant, nothing happens. Then numbers switch on their points in unison and the entire axis is activated. The coordination is faultless—the line that emerges is sharp and accurate.

You're surprised it's so simple. You see another obvious experiment to try. "Now I want you to light up only if your *first* number is 0." Just as you'd expected, the second axis lights up, its profile equally perfect.

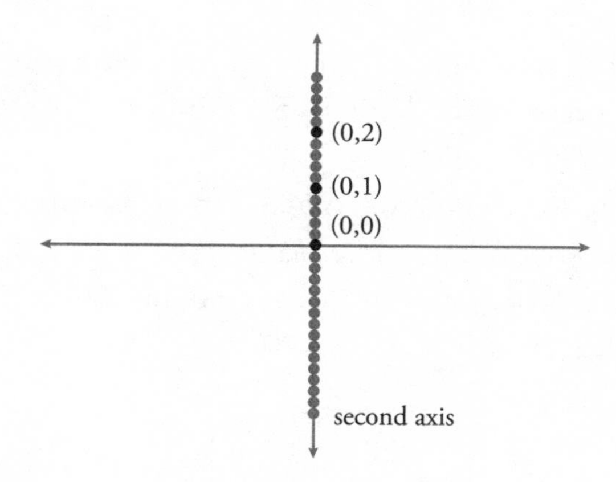

What if you change your instruction a bit, by changing the 0 to another number? With 1, you end up getting lines parallel to each axis, but exactly one unit away. Substituting larger numbers gives similar lines, but more units away.

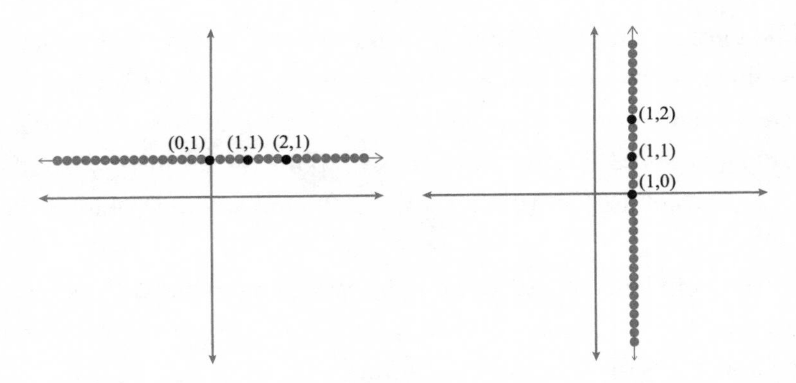

It appears, therefore, that you can generate any line parallel to either axis, without going cross-eyed trying to fine-tune which points lie on the line and should be lit. You're buoyed by this, but still stumped when it comes to the more general task. How to light up a line that tilts?

A new avatar of emptiness

The answer has its origins in games you may have played as a kid. Games that began with, "Think of a number, any number," and ended with a guessing trick.*

We're interested in only the initial step of visualization, not the trick. Imagine your mind to be a blank space. When asked to think of a number, a flock of choices flits around, one of which can alight to fill this space.

3
8
2
1 5
9 7
6 4
( )

* In one version, you're asked to double the number, then add 8 to it, then divide by 2, and subtract the original from this. The answer is always 4. Can you see why?

As a kid, you probably considered only the digits from 1 to 9. Now, however, with all the numbers we've created, you consider all the reals (we'll save the complexes for another day). So the set of possibilities flitting around now is infinite.

8.2159 $\sqrt{3}$
–4 –7.677..
7.8159 2.889 $\sqrt{21}$ 95/34
32 $\sqrt{5}$ 3/88 1/9 –46 –51.2314
6.8877 5/2 1/41
–86 2109.2
–72.93 1/35 –8.9367
1288 431
6.33333 –2 3–π –2/5 12.899
$\sqrt{92}$ 41.44 45/98
–3/4 $\sqrt{8}$ $\sqrt{7}$ 2.11

( )

Notice how nifty this is. You could never hope to enumerate this endless flock of real numbers. But just by visualizing an empty slot, you're able to transcend the finite, and generate this infinite set implicitly.

Perhaps you recognize our old friend emptiness, in an entirely new incarnation. Back when we created zero, we were dealing with the vast emptiness of nothing. Essentially, we packaged this emptiness as a number. Our new emptiness is different: a space specifically cleared with the anticipation that it will hold a real number. As long as we keep it empty, though, we're able to associate *every* real number—an infinity of possibilities—with it.

Imagine now that we partner this unfilled space with a 1. Then we get a pair of numbers, where the first number, being unspecified, could be anything.

8.2159 $\sqrt{3}$
−4 −7.677..
7.8159 2.889 $\sqrt{21}$ 95/34
32 −46 −51.2314
$\sqrt{5}$ 3/88 1/9 6.8877 5/2 1/41
−86 2109.2
−72.93 1/35 −8.9367 431
1288
6.33333 −2 3−π −2/5 12.899
$\sqrt{92}$ 41.44 45/98 $\sqrt{7}$ 2.11
−3/4 $\sqrt{8}$

(,1)

So this pair could include (−4,1), (−3/4,1), (0.5,1), (2.5,1), (3.6,1), and so on, that is, every real number matched with 1. Which means that to represent (,1) graphically, we should include all the corresponding points, in other words, an entire line.

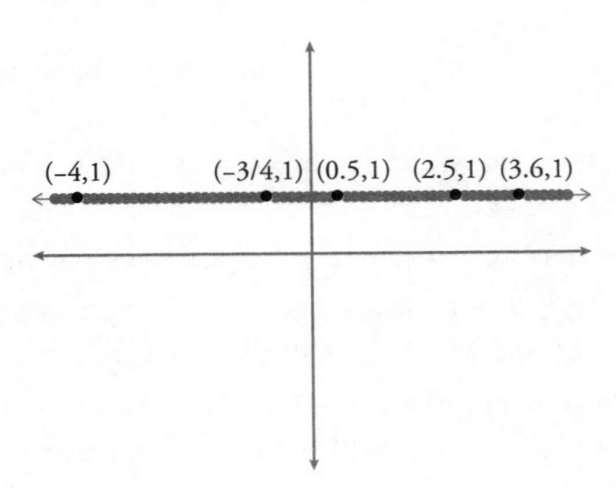

"Light up if you're a pair of the form (,1)" is, then, the instruction equivalent to "light up if your second number is 1." The key is the empty space. Since emptiness doesn't have a commanding-enough presence (one might not even know it's there), let's replace it with the symbol we've all been waiting for, the number one poster child of abstraction: *x*.

There we have it—the familiar, beloved, and dreaded *x*. The gateway to algebra, and with it, much schoolkid anguish. I once saw a statement on an "I hate algebra" website that went, "Numbers and letters are like gasoline and milk, they should never be mixed."

But replacing numbers with letters is actually an astonishing idea, one that liberates us from much numerical drudgery. Because algebra isn't just about solving equations for x (as they too often make you do repeatedly in school, without explaining why). Beyond its computational capabilities, it's also shorthand—the closest mathematics gets to a language. It allows you to express a relationship that might apply in thousands of instances, an infinite number of them. For instance, imagine if the only way to specify a line was to enumerate every point on it, coordinate by coordinate. You'd have to say what happened not only when the first coordinate was 1 or 2, but also when it was 1.1, 1.01, 1.001, and so on. Not to mention all the irrational values in between as well. The task would be impossible for any human, because of the infinite number of points involved. It's only with x at our command that we can, in one concise swoop, accomplish this task. For instance, by using $(x,1)$ to generate the line above.

Putting a slant on it

So how to create a line that's slanted with respect to the axes? For this, let me propose the following guessing game. "Think of any number and I'll tell you what its double is."

It would be futile for me to try to win this game by actually attempting to guess the numbers you mentally imagine—I'd be very bad at it. But here's my trick: using x, I can tell you the double even before you choose the number. I'll simply represent any number you decide on as x, and then furnish you with its double, which has to be $2x$. So there—I've doubled your number without knowing what it was!*

You may think that's glib, but I'm serious; x provides the ability

* As a simple extension, here's how using x readily explains the kid's guessing trick mentioned in the previous footnote. Denote the number you've thought of as x. Then doubling it gives $2x$, adding 8 gives $2x + 8$, dividing the answer by 2 gives $x + 4$, and subtracting x always gives 4.

to double every number that exists. One no longer has to enumerate each possibility, only write down the answer $2x$. Just like with the line $(x,1)$, it's a way to manage the infinite.

Let's take a look at that line again. The instruction $(x,1)$ tells you to go x units along the first axis (call it the horizontal direction), and then 1 unit perpendicular to this (call it the vertical direction). That's what gives rise to the line.

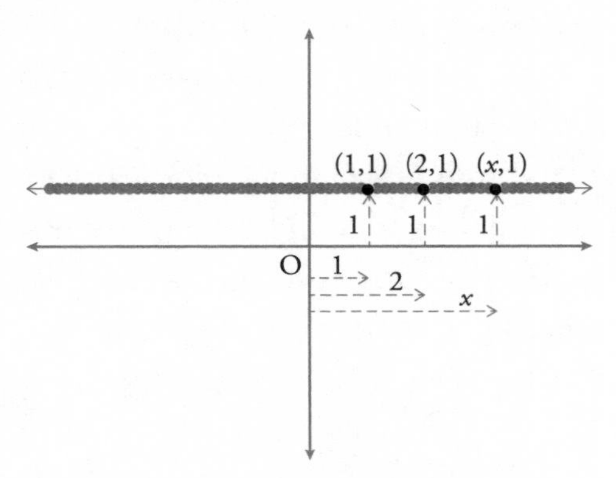

What if we change this to $(x,2x)$? This would mean go x units in the horizontal direction, then twice the same number of units vertically. With 1, you end up at the point (1,2), with 2 at the point (2,4), with 1.5 at (1.5,3), and so on. In other words, the point x is raised to a height $2x$, a rule that we'll also represent as $x \rightarrow 2x$. Notice what emerges.

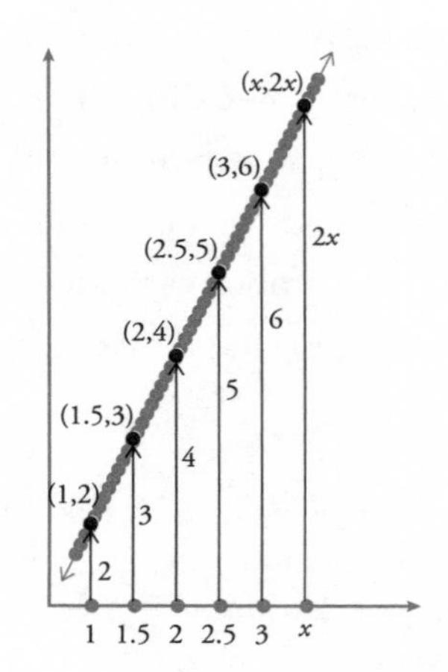

Yes, it's a slanted line. Every point on it has coordinates of the form $(x,2x)$. In fact, asking all pairs with this form to light up will make this line spring into life, all glowing and connected. You've done it!

Now you can not only show Nature what you mean by a straight line but also teach her to draw a bunch of them. All she needs is to specify the formula like $(1,x)$ or $(x,2x)$, so the numbers know which points to illuminate. You're beginning to figure out how to communicate—these formulas are part of the starter kit for your language.

17.

THE WORLD OF POLYNOMIALS

IN THE INTRODUCTION, I MENTIONED I WOULD NEED to "severely limit" formulas and equations, to ensure retaining all my readers. A critical issue indeed, now that we're in algebraic territory and child-eating symbols like x roam the land. In fact, there's an old adage cautioning museum curators that each equation they add to a science exhibit will drive away half their remaining audience. It's a warning I should surely heed. What if you—or worse, the pope—gets turned off?

Actually, he's one person I don't have to worry about. The pope, you see, has studied more algebra than you might suspect. He graduated from a technical school and—look it up—even worked in a lab as a chemical technician. The glimpses of x so far have been a welcome break from the weighty issues he's constantly grappling with. He's curled up now with an advance copy of this manuscript (which my publishers sent him with a request for a blurb), all cozy in bed. Picture him in his papal pajamas, reacquainting himself with his old math friends.

Will the rest of my readers take inspiration from the pope? I know some might just tune out when they see the x, figuring anything with symbols is beyond them. Well, one aim of this book is to prove that's not the case. So I'm going to stretch the boundary a bit regarding the pledge I made. I'll be including just enough formulas here so that you don't feel intimidated by them again.

Why do we need formulas anyway? Recall that we're trying to find a language to communicate with Nature, one that can describe the components of the universe to come. We've had a successful start, creating the symbol x and showing Nature how to generate straight lines with it. Now we want to develop this language further, by coming up with formulas to fashion more shapes. Not just standard ones like circles, but all the profiles Nature's going to need in her repertoire. Imagine a profusion of geometric forms—fruits, flowers, plants, animals, planets, and more—all bursting with curves.

There's a problem. Formulas are terse and inscrutable, and don't obviously conjure up the lush elegance of flora or fauna. How does one look at a particular equation, with its assortment of x's, and get any clue of the shape it's supposed to evoke?

While this is hard to do in general, there's a basic class of formulas, which proves to be the most versatile and useful, that is easier to get a handle on. In the next few sections (which more math-savvy readers can quickly skim) we'll see what these bread-and-butter formulas, called *polynomials*, are all about.

The simplest case: Linears

Let's start with better understanding the straight line $(x,2x)$ we just drew. Many of you must have encountered a different formula for it in school, so let's switch to that. This involves introducing a new symbol y that denotes the vertical height to which each point x along the horizontal axis is raised. Rename the two axes the *x axis* and the *y axis*—labels we have Descartes to thank for. Then, instead of $(x,2x)$, we can write (x,y), and denote the line by the more familiar equation $y = 2x$.

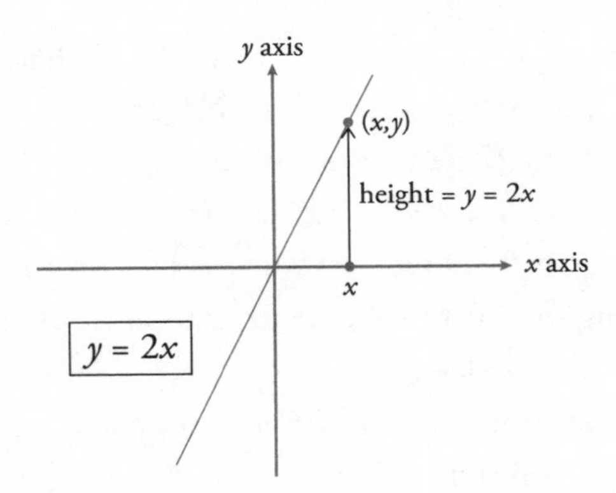

There's nothing special about the 2 in this formula. One could equally well take 3 to draw the line $y = 3x$ (which would be steeper), or take $\frac{1}{2}$ to get the line $y = \frac{1}{2}x$ (which would be *less* steep). This number—called the *slope* of the line—is an indication of how much the line tilts. Here's a sketch to show which lines correspond to different slopes (which could be zero or even negative).

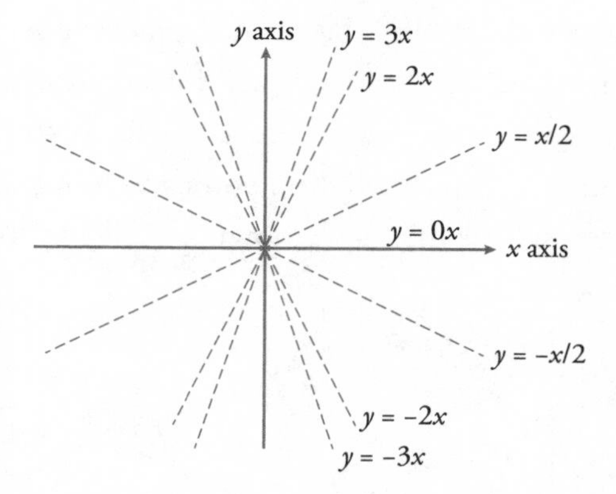

There's a second aspect to straight lines besides the slope. Let's say we consider the equation $y = 2x + 1$. Plotting this gives a line parallel to $y = 2x$, since the +1 simply moves every point up by one unit. Changing this +1 to +2 or −2 or some other number gives a series of parallel lines. This number, called the *intercept*, measures where the line cuts the y axis.

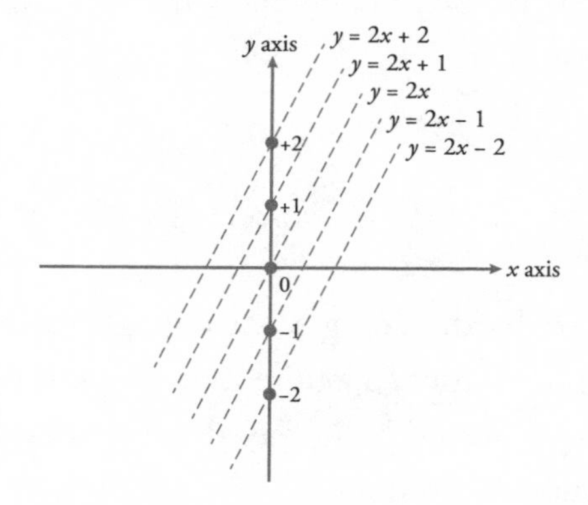

The line is actually a very cut-and-dried construct—it's governed entirely by these two numbers, the slope and the intercept. In fact, we can write the equation of any straight line (or *linear*, as we will call it) as

$$y = (\text{slope})x + (\text{intercept}) = (\text{first real number})x + (\text{second real number}).$$

These two numbers are like the control knobs of a radio—you can use them to tune in any "station" (i.e., linear). Adjust the first knob (i.e., the first real number) and you can match the tilt of the line you're after. Adjust the second, and you can complete the zeroing-in process, by moving the line up or down, parallel to itself.

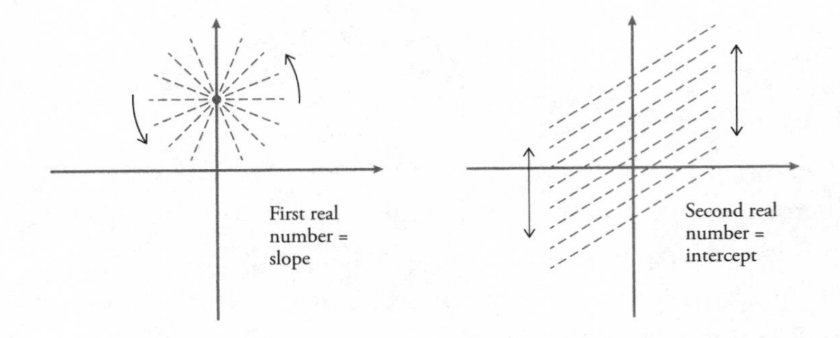

You can now teach Nature to generate any line she wants, simply by specifying its slope and intercept.* Once she masters the line, she can use it to draw the boundaries of triangles, squares, rectangles—in fact, any polygon.

The exponent game

Straight lines are hardly going to be enough to generate the profusion of shapes we mentioned before. Imagine a universe built entirely of such linears—it would be just as alarming as one populated with droves of inscrutable formulas!

---

* With some more algebra, which we won't do here, you can also teach her a different formula, with which she can draw the line joining any two specified points.

But how is Nature going to generate curves? Our hope is the answer will lie in the symbols we've created. If these symbols can generate straight lines, why not other shapes?

To proceed, it's helpful to think of x being imbued by a little bit of synesthetic magic. After all, the symbol is a stand-in for numbers, so why shouldn't it be able to play their games? In fact, you can think of all the linears we've formed so far as simply arising out of such play: first x multiplies a number (as in $2x$ or $3x$) and then plays the addition (or subtraction) game with another number (to get formulas like $2x + 1$ or $3x - 2$). The problem is that no matter what combinations you create by making x play these addition, subtraction, and multiplication games, you succeed only in generating more linear formulas, that is, more straight lines.* How to escape this limitation?

The answer comes from a game Two suggests to x. "Instead of multiplying yourself by a number, why not try multiplying yourself by yourself?" There's an ulterior motive Two has—to insert himself into this game as a symbol attached to x. Just as $x + x$ is denoted by $2x$, he proposes they denote $x \times x$ by x^2.

* Notice that multiple games will still result in just linear formulas. For instance, $(2x + 1) + (3x + 2) + (4x + 3)$ just gives $9x + 6$. Also notice that we have pointedly excluded the game of division by x, which we save for Day 5.

In short order, Three has convinced x to form $x \times x \times x$, and denote it by x^3, and Four has done the same with x^4. Pretty soon, every natural is inviting x to play this "power" game. Eager to explore this synesthetic world, x readily obliges.

Naturally, you get very curious about these powers. What kind of graphs will they generate? As a test, you set up the equation $y = x^2$ and ask points to light up according to it. The result is a strange new curve never seen before in your universe. To your delight, this *parabola* (as you call it) has broken through the barrier of the straight line!

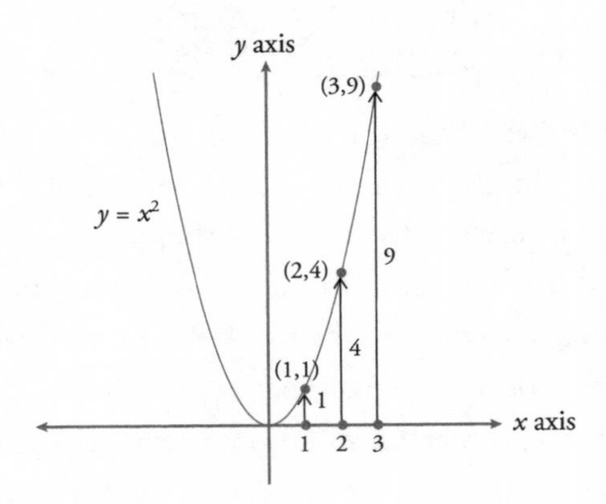

How did this happen? Each value of x has been raised to a height x^2, so that 1 just makes it to height 1, 2 is elevated to height 4, while 10 is elevated to a whopping 100! These heights are no longer proportional to x, as they would have been under the doubling rule $x \longrightarrow 2x$. You realize that the inclusion of a squared term like x^2 ensures the graph can no longer be straight.

In fact, the presence of any other power (x^3, x^4, x^5, . . .) also leads to a curve. You notice that as x increases, the lines generated by $y = x^3$, $y = x^4$, $y = x^5$, and so on, grow faster than any of their predecessors in the list. Their graphs look like strings leading up to invisible kites, each flying higher than the previous one.

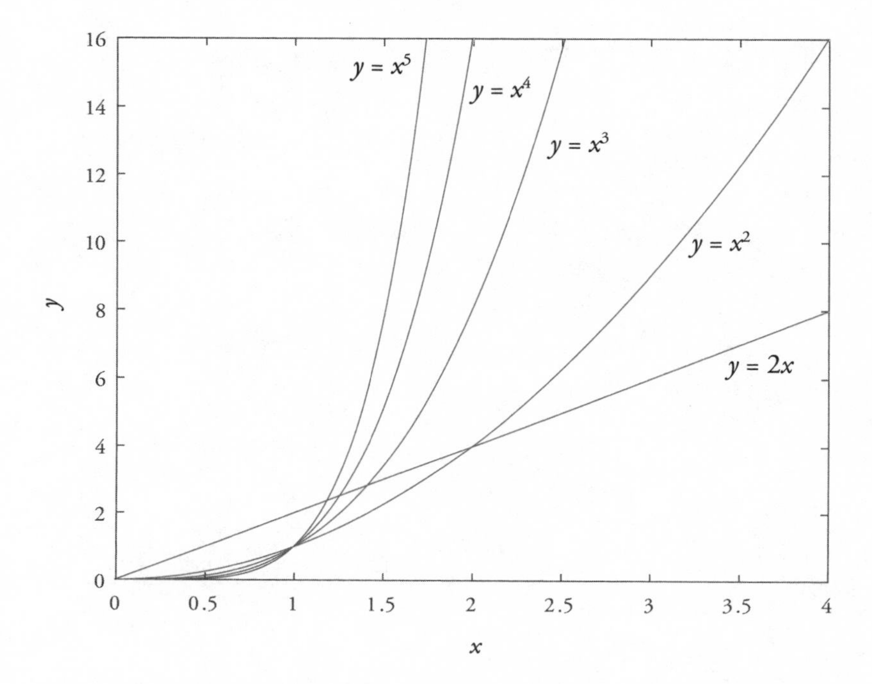

Nature quickly assimilates these new possibilities. But a tall barrier looms up ahead. She still can't draw the curve that's going to be most essential to your universe: the circle.

Roller-coaster formulas

The circle is the ultimate challenge. Generating it is a lot harder than you expect. The powers of x all zoom upward rather than curving back on themselves the way a circle does, so none of them work. You play around with x^2, multiplying it by reals, adding other reals to it, even coming up with formulas that incorporate x as well, like $2x^2 - 7x + 1$ and $-4x^2 + 5x - 5$. But the graphs of such expressions turn out to be just more parabolas—moved around, turned upside down, narrower or more open, but parabolas all the same.

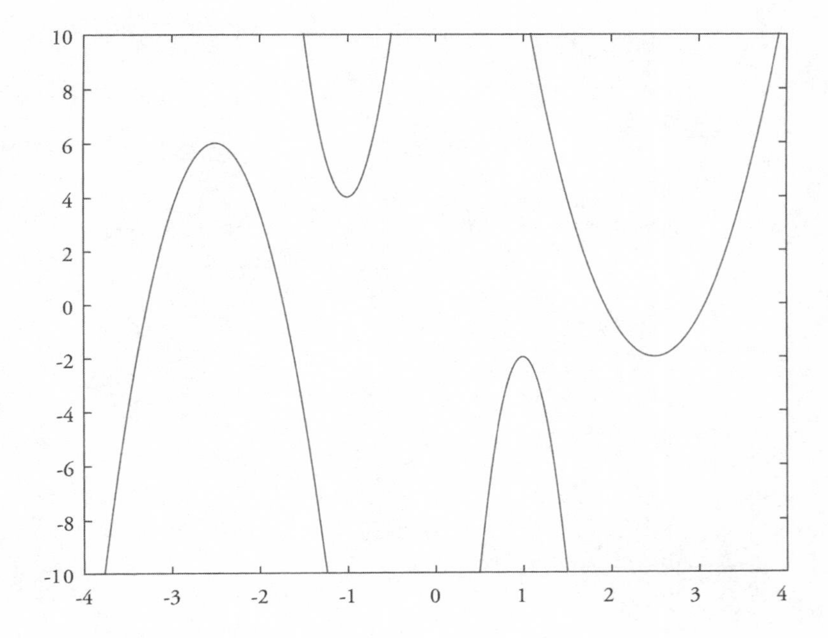

Let's try to understand the formulas we've generated so far. First, the *linears* obtained using x were of the form:

$$(\text{first real number})x + (\text{second real number}).$$

Now we've added x^2 to the mix, to get new formulas called *quadratics*:

$$(\text{first real number})x^2 + (\text{second real number})x + (\text{third real number}).$$

The graphs of linears, you've seen, generate straight lines, while those of quadratics generate parabolas (which, unlike straight lines, have a hump or mound—either a low point or a high point).

What if we create more such formulas? For instance, adding x^3 into the mix gives *cubics*:

$$(\text{first real number})x^3 + (\text{second real number})x^2 + \\ (\text{third real number})x + (\text{fourth real number}).$$

What would the graphs of such cubics (e.g., of $y = 3x^3 - 4x^2 + 2x + 1$) look like?

Drawing a few cubics shows they often have an interesting feature that can distinguish them from quadratics. Whereas a quadratic has a single hump, a cubic often has two—one low and the other high, relative to everything around.

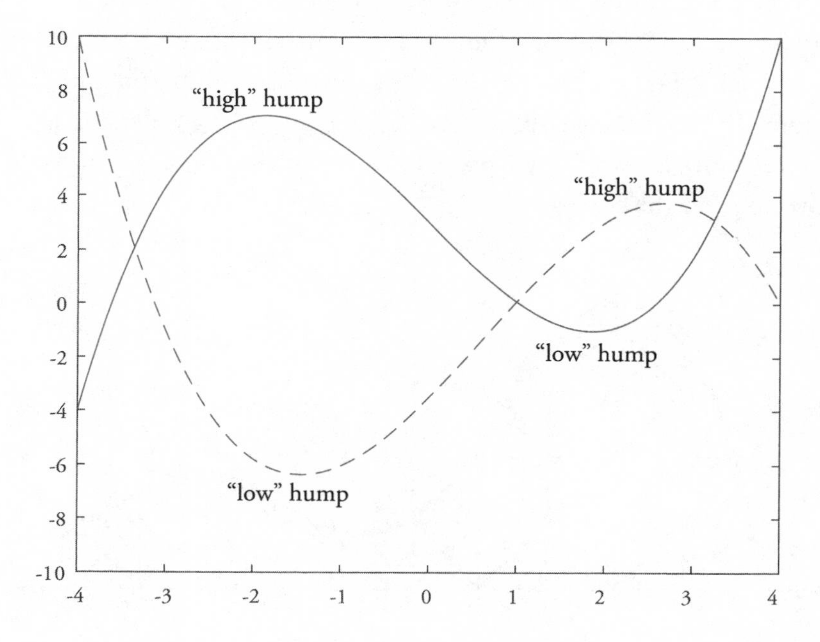

A good way to think of the real numbers in a cubic formula is in terms of control knobs again:

$$\text{cubic} = (\text{first knob})x^3 + (\text{second knob})x^2 + (\text{third knob})x + (\text{fourth knob}).$$

Recall that a linear had two control knobs: the slope and the intercept. Adjusting the first one tilted the line, the second made it move up and down. Now, for a cubic, we have *four* knobs controlling various aspects: the height of the humps, their position, their width, and so on. With these, we can make the graph go up or down, open up, flip around. It's not as clear-cut as in the linear case, so you need more know-how. But with enough practice operating the knobs, you can tune the cubic into a range of different shapes.*

What if you include more terms (like x^4, x^5, x^6, . . .), each with its own knob? Then the resulting combination, called a *polynomial*, can have even more humps—three of them when an x^4 term is added (these are called *quartics*), four with x^5, ninety-nine with x^{100}. More knobs give you increasing flexibility to make the graph undulate more. Think of the resulting curve as the design of a roller coaster—by adjusting the knobs, you can control both the number of peaks in the ride and their steepness.

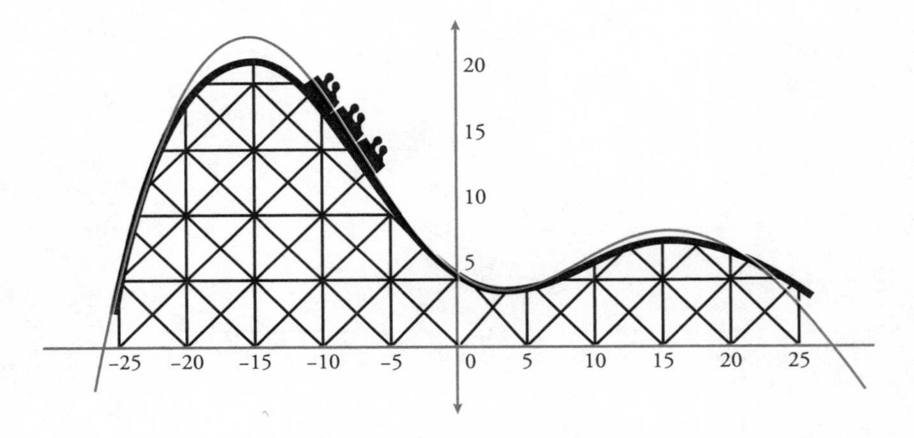

While Nature's not going to be designing any roller coasters soon, you can see now how evocative polynomials are, how they can inspire all sorts of natural forms.

* You can play around with these knobs to see how they work on an online site called Desmos, in the case of both quadratics (https://www.desmos.com/calculator/zuaqvcvpbz) and cubics (https://www.desmos.com/calculator/wqufjhhj8y).

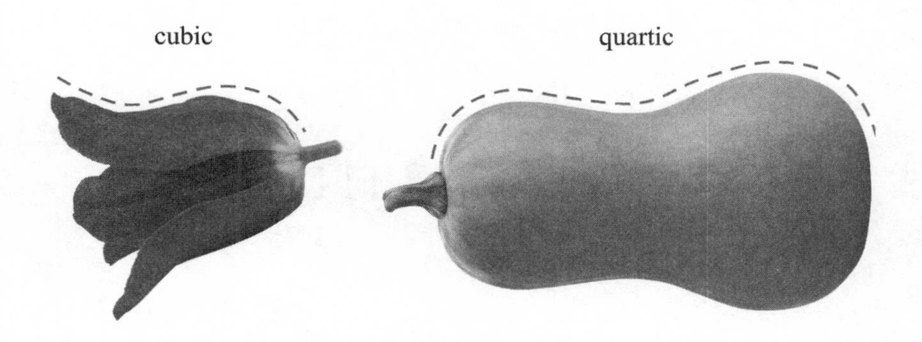

That's all we're going to need to know about polynomials for now. They generate shapes you might want, and are able to wiggle an extra time with each power of x included. Keep this in mind and you need never be intimidated again when confronted by such formulas.

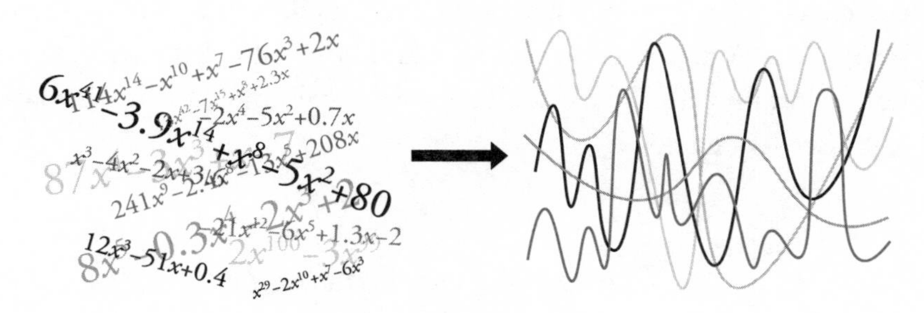

Despite all their flexibility, one thing the above polynomials *can't* do is curl around to join back up. In terms of roller coasters, this means that a "loop the loop," where riders are turned upside down and back again, cannot be generated. Nor can such polynomials yield something as basic as a circle. Something new is needed for this, something more than just the powers of x.

18.

THE *y* OF THINGS

YOU MIGHT HAVE NOTICED THAT y, SO FAR, HAS BEEN hanging around on the left side of the "equal to" sign in equations, without being deployed in interesting ways as x has. To win the circle challenge, you may need to include its powers as well—surely some formula involving x and y should do the trick. Let's see how to find such a formula. This will also give us insight into how mathematical discovery works.

The experimental method

Historically, there have been two main tracks humans have taken to come up with new mathematical truths. One is deduction—start with what's known (or assumed) and then build up results through logical reasoning and inference. The resulting proof, if correct, is a validation forever—it's also available for anyone to check, in order to convince themselves. The ancient Greeks, who valued pure mental exploration over any other kind, spearheaded this approach. A prime example is Euclid's monumental work of building up geometry from basic assumptions, which we paid homage to on Day 2. To keep things brisk, we haven't included proofs in the main text, but several examples can be found in the endnotes.

In contrast to proof-based deduction, much of the basic mathematics we use today arose through experiments. It was distilled from the experience accumulated by the world's earliest builders and architects, its engineers and navigators, scientists and astronomers. Mathematical rules were a way of interpreting past phenomena to make future predictions; they often evolved from centuries of observation and experiments, of trial and error. As an example, although Pythagoras may (or may not) have proved the famous theorem that bears his name, such properties of right-angled triangles had been observed centuries earlier, in lands as far-flung as Babylonia and India.

To figure out a formula for the circle, we're going to throw in our lot with the experimentalists and tinkerers. Except we have no building materials, nothing physical to experience. No matter—we'll use numbers to perform our experiments, pore over them to see if we can detect a pattern (mathematics, incidentally, is often called the study of patterns). Specifically, we'll examine a bunch of points on the circumference of a unit circle (a circle with radius 1) and see what their x and y values tell us.

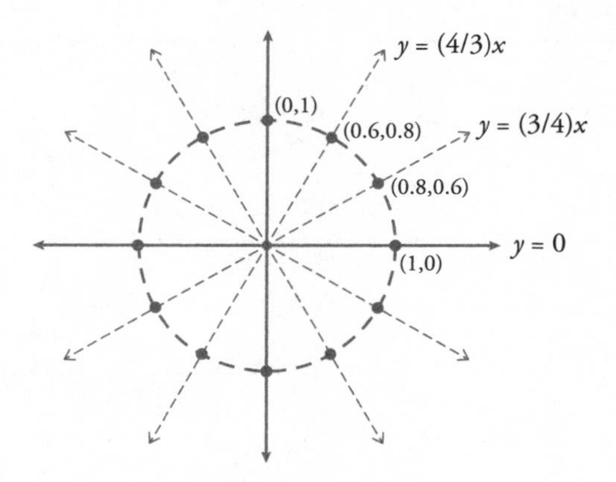

How do we identify such points? We know these all lie a distance 1 away from the center (taken to be the origin). If we draw any straight line through the origin, we can then walk one unit along this line and note the condo unit at which we arrive. For instance, proceeding along the x axis puts us at (1,0). Walking one unit along the line $y = (3/4)x$ brings us to (0.8,0.6), while the line $y = (4/3)x$ deposits us at (0.6,0.8). These coordinates might seem to come out of the blue, but they're obtained by simple measuring (the "experiment" part). We can also proceed one unit in the opposite direction along these lines, or along the lines $y = -(3/4)x$, $y = -(4/3)x$, and others, to get more points.

Once we have a bunch of (x,y) values for points on the circle, the second part of our experiment begins. We list various formulas involving x and y, and see if they tell us anything interesting. We're not sure quite what to expect or look for—but we're hoping for something to stand out, as it must have for the ancient Babylonians when they were playing around with right-angled triangles. In the following table, we've calculated the values of $x + y$, $x - y$, x^2, y^2, $x^2 + y^2$, $x^2 - y^2$ for the twelve points shown in the preceding diagram. Our idea is that if we don't see anything striking, we'll continue by computing other formulas, like x^3, y^3, $x^3 + y^3$ and so on, that is, keep fishing around until we find something.

(x,y)	$x+y$	$x-y$	x^2	y^2	x^2+y^2	x^2-y^2
(1,0)	1	1	1	0	1	1
(0.8,0.6)	1.4	0.2	0.64	0.36	1	0.28
(0.6,0.8)	1.4	−0.2	0.36	0.64	1	−0.28
(0,1)	1	−1	0	1	1	−1
(−0.6,0.8)	0.2	−1.4	0.36	0.64	1	−0.28
(−0.8,0.6)	−0.2	−1.4	0.64	0.36	1	0.28
(−1,0)	−1	−1	1	0	1	1
(−0.8,−0.6)	−1.4	−0.2	0.64	0.36	1	0.28
(−0.6,−0.8)	−1.4	0.2	0.36	0.64	1	−0.28
(0,−1)	−1	1	0	1	1	−1
(0.6,−0.8)	−0.2	1.4	0.36	0.64	1	−0.28
(0.8,−0.6)	0.2	1.4	0.64	0.36	1	0.28

Glance at the table and you'll see we've hit pay dirt! While other formulas assume a bunch of different values, the $x^2 + y^2$ column springs out, since it's always exactly 1. Could this be a coincidence? Perhaps we've fortuitously chosen exactly those points for which the sum of x^2 and y^2 is 1, and there's a vast majority of points lurking on the circle for which this isn't the case? So go ahead, experiment some more, by locating a bunch of other points on the circle and calculating $x^2 + y^2$ for them as well. You'll find the results unequivocal. The sum is the same for every single point you try!

You stare at all the 1s you've calculated. Do they contain the answer to the riddle? Could the equation of the circle be what it seems? That the sum of x^2 and y^2 equals 1?

It's time to test this hunch. You instruct only those points to light up whose x and y satisfy

$$x^2 + y^2 = 1.$$

Instantly, the unit circle springs to life. The experimental method proves itself!

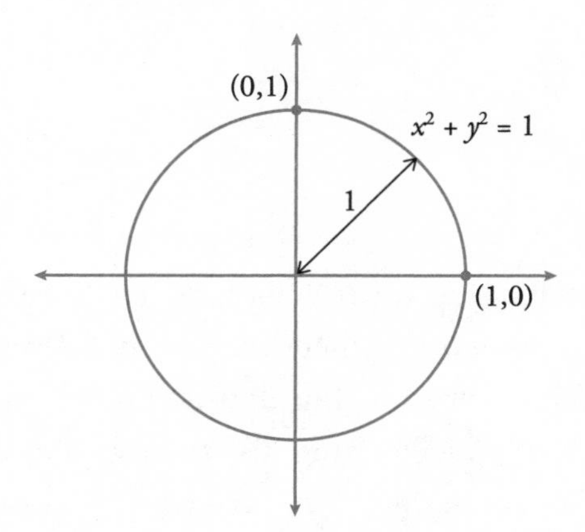

You soon find that a circle with any radius can be drawn by changing your equation to

$$x^2 + y^2 = (\text{radius})^2.$$

A straightforward modification to the formula (which we won't get into here) also allows you to change the center to any other point.

Nature seems to appreciate the circle much more than the triangles and rectangles she was able to draw with straight lines. She generates a bunch of circles, gleefully calling out their equations.

Teaching Nature how to draw

Once y has been squared, you can introduce other powers of it into equations as well. In fact, you can even multiply together powers of x and y. The expanded family of polynomials that results (including such terms as xy, x^2y, xy^2, x^2y^2, x^3y, and so on) leads to the equations of some intriguing new shapes being discovered.

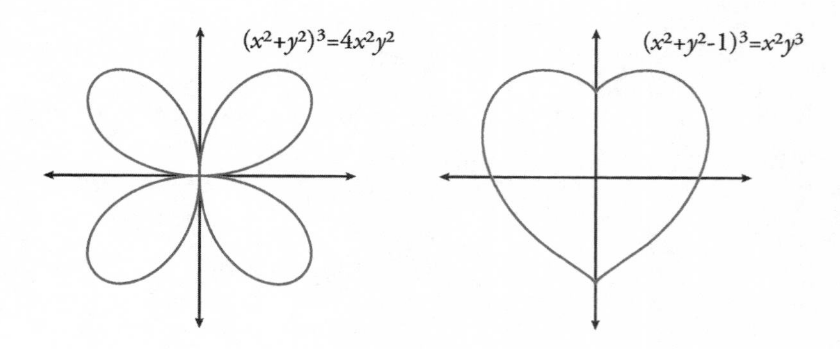

You could try plotting other equations involving x and y to discover even more shapes, but there are two problems. First, Nature will have a hard time remembering all these exotic equations. (In fact, you will as well—try memorizing just the two above, as a sample.) Second, it's not clear that any arbitrary curve you draw will always have an equation—and even if it does, whether you'll be able to find this equation. So developing such formulas may not be the way to ensure Nature's able to generate all the shapes needed in the universe.

Fortunately, there's a better strategy. It's possible to approximate any shape desired just by marking off a few points along the boundary and joining these with linear segments. For instance, a pretty good likeness of a heart can be generated by connecting just eight points along its boundary (as shown in the picture on the left).

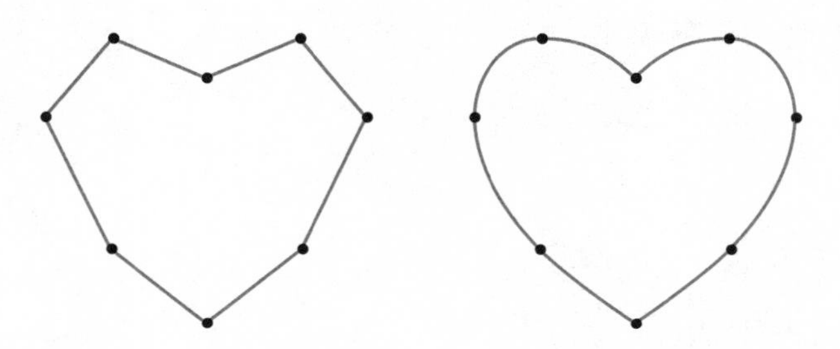

But such piecewise outlines don't have the smoothness of the original. By using the smoother *cubics* instead of linears for each piece (as shown on the right), you can essentially eliminate all corners and kinks!*

Such "piecewise" methods free both you and Nature to generate a variety of shapes. No longer does either of you have to remember a manual of complicated equations. Instead, all you need to do is specify some points, and use linear or cubic connections to generate a shape (much as specifying two points allows you to generate a straight line). For any outline you have in mind, just transfer to Nature enough points lying on it and she'll fill it in. What could be easier?

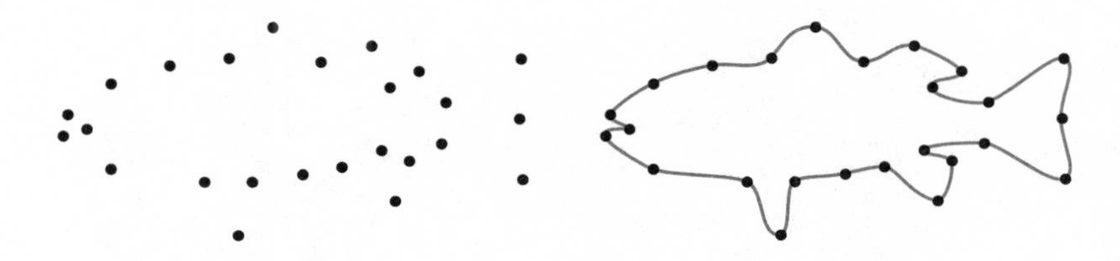

* Also, by using enough points, you can get as close to the original shape as wanted (see the endnotes for more).

You practice a few such transfers with Nature, and they all go smoothly. The gates of geometry have been thrown open. Nature can now generate all the curves you'll need her to, using just the algebra of simple polynomials. You encourage her to experiment, create some new shapes herself, to get the hang of the process.

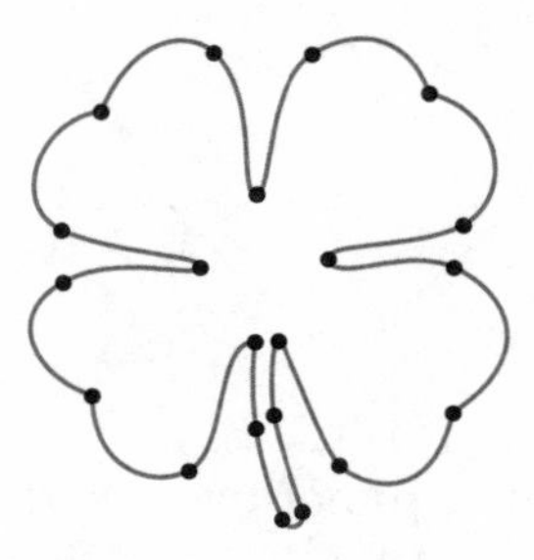

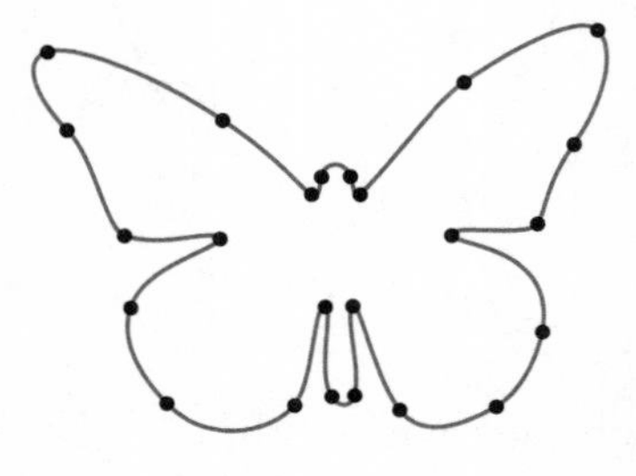

19.

THE ALGEBRA OF DISTANCE

NOW THAT WE'VE SET UP FORMULAS TO DESCRIBE different shapes, let's see what else this language may be good for. You're probably anticipating that I'll bring up the "real" use of algebra they taught you in school: solving equations. Surely that should be essential while designing the universe. All the slogging you put in back then will finally pay off. Even the pope's itching to test whether he can still recite the quadratic formula.

It's a handy skill, but one we won't saddle Nature with just yet.* She has a more pressing need right now: how to measure distance. As we'll see ahead, algebra provides the answer. However, it provides *several* possible answers, in other words, several plausible formulas for distance. Not only does this mean that we can't take a fundamental concept like distance for granted, but it also challenges our understanding of a basic geometric shape: the circle.

---

* Let me point out one advance we've made over high school. If you recall, some quadratic equations, like $x^2 + 1 = 0$, were said to have no solutions. Now, however, we can verify that both $x = i$ and $x = -i$ are solutions. In fact, with our introduction of complex numbers, *every* polynomial equation in x has solutions; linears have one, quadratics have two, cubics have three, and so on (with repeated solutions being counted individually).

The distance formula

Recall that so far, we've depended on physical measurement of distance, counting out the units as one walks along a straight line between two points. We'd now like something easier. Given a point (x,y), we'd like a formula that tells us how far the point is from the origin.

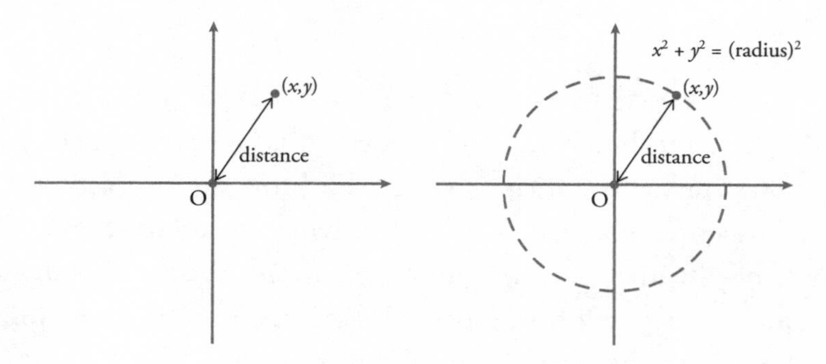

The answer lies in the equation of a circle we recently discovered. Imagine a circle drawn through (x,y) with its center at the origin. Then our distance becomes the circle's radius. Substituting radius = distance, the equation of the circle can now be written as

$$x^2 + y^2 = (\text{distance})^2.$$

Taking square roots on both sides then gives the formula we want:

$$\sqrt{x^2 + y^2} = \text{distance}.$$

Nature no longer has to don her hiking boots each time she wants to measure such a distance. All she need do now is plug x and y into the above.

Although we arrived at this distance formula through observation, it can also be established through proof. Complete the right-angled triangle with x and y as sides, and use the symbol d for distance. You'll notice something familiar. The distance formula has just become the classic Pythagorean theorem: the square of the hypotenuse is equal to the sum of the squares of the other two sides. A result that was mathematically proved several centuries ago!

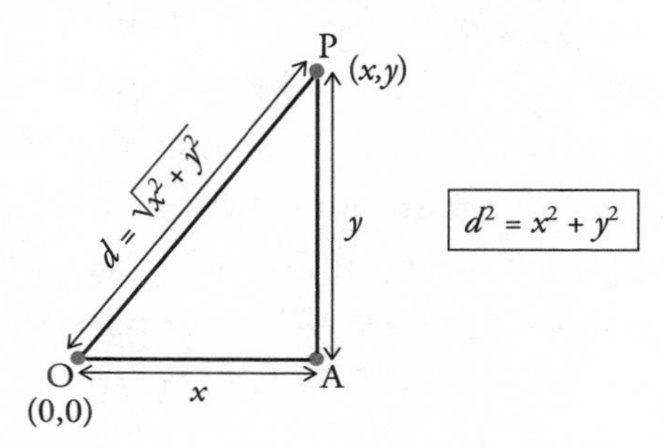

This is another ability afforded to us by symbols and algebra: they allow us to express such immutable truths. For instance, the guarantee that any right-angled triangle we construct in our universe will satisfy Pythagoras's theorem. There's a caveat, though—the underlying space we choose for our universe has to be flat. Introduce curvature into it (a possibility we explored on Day 2) and both the distance formula and the right-triangle relationship will have to be modified. *You* get to decide what universe you want—one that concurs with Pythagoras or not.

Squaring the circle

Surprisingly, even if your space is flat, you can choose among different distance formulas for your universe. As I mentioned, there's more than one way to measure how far points or objects are from each other. The time has come to show you one of these alternatives: a different way to quantify "farness."

Consider again the above diagram of the triangle. The usual way to measure distance is along the hypotenuse, but what if you went along the two sides instead? Specifically, rather than $\sqrt{x^2 + y^2}$, what if you defined the distance from O to P as $x + y$? Since distance is a way of ascribing a relative value to how far one point is from another, why shouldn't this be equally valid?

But this isn't the *real* distance, you object. Not according to how the crow flies.

You're right, it isn't. Then again, we're not crows. Suppose you build your universe complete with cities and roads and cars and humans. Then for two urban locations O and P, we might be more likely to be traveling in a taxicab, rather than along the straight line of a mythical crow. Say all the roads are arranged in a grid, as can be typical in a city, with no diagonals to cut across. What distance would be relevant then? The one your taxi fare would be based on?

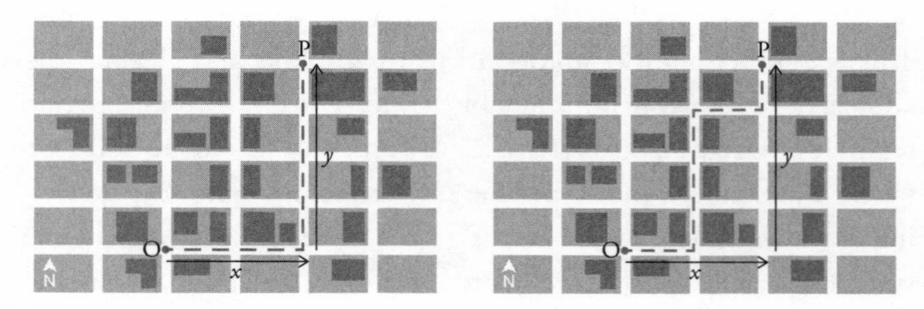

The answer, clearly, is $x + y$, not $\sqrt{x^2 + y^2}$. In fact, if you take some zigzag route, as shown above on the right, then adding the separate pieces might still give the same total distance (and the same fare). This definition seems much more appropriate—certainly as far as the bottom line of your wallet is concerned.

There's an apt name for this alternative formulation of distance: the "Taxicab" or "Manhattan" metric. It illustrates not only that there are different ways to assign useful meaning to the concept of distance but also that you have to design the right definition, the right mathematics, for each application.*

Here's an interesting question. Suppose this taxicab distance to

* As an example of a variation, the Washington, DC, metro system bases its fares between two stations on the *average* between the actual distance traveled along the subway lines and the distance between them as the crow flies.

your point P is 1 mile, resulting in a fare of ten dollars. To which other points could you be transported for exactly the same fare, assuming all roads are on the same perpendicular grid? In other words, which other addresses, when measured "as the taxicab drives" instead of "as the crow flies," are exactly 1 mile away?

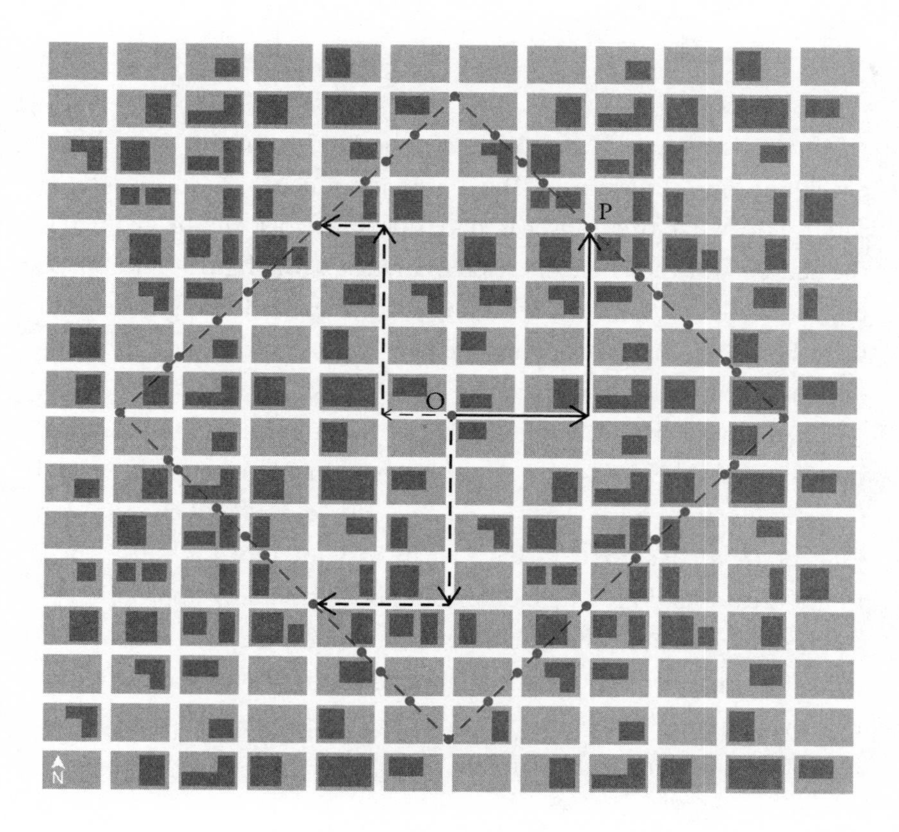

Try out routes to different points, and you can verify that such addresses lie along a diamond shape, whose center is the starting point. But wait, don't we already have a word that describes such a figure? A figure obtained by plotting all points the same fixed distance away from a center? We do indeed—it's called a circle! Which, surprisingly enough, is exactly what you're looking at: a circle with radius 1 mile and center at O, but using this alternative definition of distance instead.

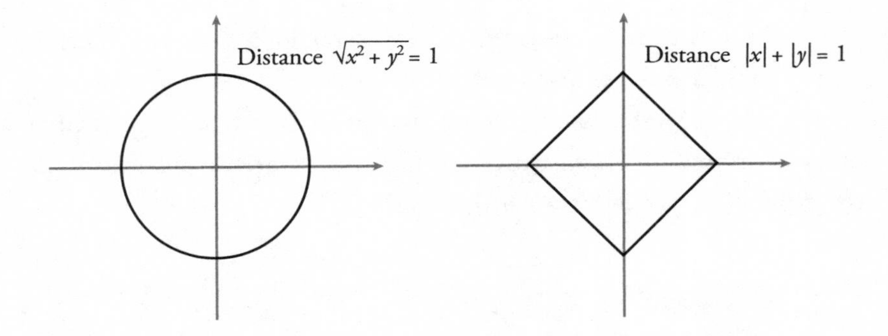

Above, I've plotted two "circles," obtained from our two different definitions of distance.* The diamond, of course, is just a rotated square. So what this shows is remarkable. Two icons of geometry, the square and the circle, are essentially the same figure—the shape depends only on the way one defines distance! Something to remind us of how even our most commonplace labels can depend on underlying mathematical assumptions.

Let me remark that there are many other formulas for distance that could also be used—for example, those involving cube roots and fourth roots (see endnotes). The shape of the "circle" will change with each of them. It's good to have them in your tool kit, since there might arise situations where one would be particularly appropriate.

For now, though, we'll measure everything using our normal distance formula $\sqrt{x^2 + y^2}$, imposed as the default. Nature's already worried that you might now expect her to fit a square peg in a round hole. It's best not to perturb her further.

Going forward

So what have we accomplished with algebra? We've developed a bunch of tools for our contractor, Nature. Lines, circles, polynomials, and

* For the taxicab formula, $|x|$ and $|y|$ indicate absolute values, i.e., if x or y is negative, we change the sign and input the corresponding positive value instead. This ensures that the distance never becomes negative.

piecewise curves with which she can generate various shapes and figures in a coordinate system. (The formulas will also come in handy to express the laws of physics, as we shall see later.) We've also found a precise and convenient formula for distance, so that Nature can process all the measurements we give her. Overall, we've created a language with which to communicate our instructions. What we want, what specifications to follow—in other words, all the information one must give a contractor.

There's one gap that remains, so let's fill it now. So far, we've developed this language only in the simplified context of the plane formed by the first two axes. Consequently, all our shapes have been in 2-D. How do we extend our work to 3-D?

Some shapes lend themselves to a quick and easy trick. Just rotate them around a central axis, and the resulting surface traced out will be a natural 3-D extension.

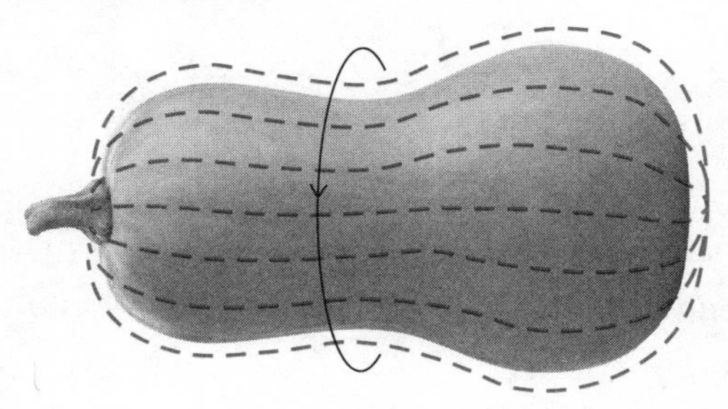

To handle more general cases (including those that don't have symmetry around an axis), let's add a third symbol z that measures distance along the axis perpendicular to the 2-D plane we were working in. In other words, we specify any point in 3-D by the coordinates (x,y,z).

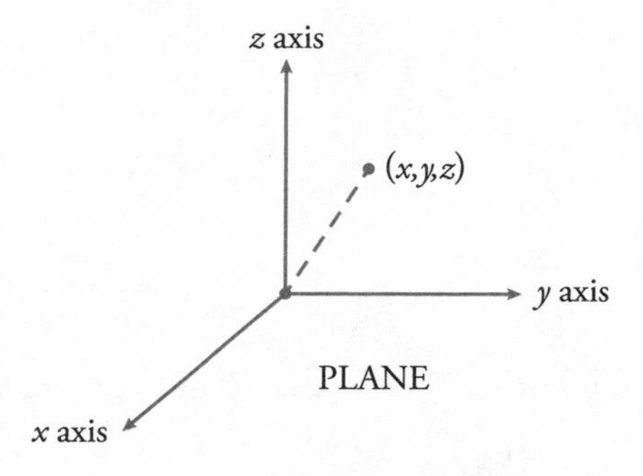

Lines between pairs of points in 3-D can now be described via equations between x, y, and z. We won't dwell here on the details. Suffice it to say, you teach Nature all the basics needed. This includes how to construct arbitrary solids—instead of specifying bounding lines, she must now specify bounding faces. In particular, by using facets (rather than piecewise lines) to join a bunch of points, she can generate arbitrary shapes.

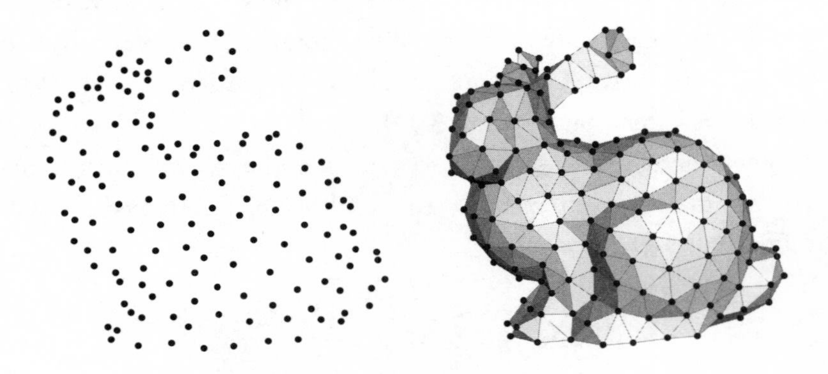

As in 2-D, using more points for the shape she has in mind can give a closer rendering, while using the analogs of cubic polynomials can create a smoother effect.

Nature's particularly taken by the sphere (which she learns to generate by a simple modification of the circle's equation, which now involves z as well).

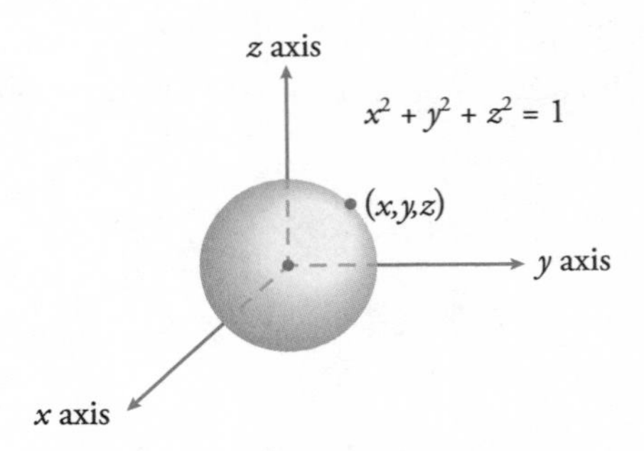

It's probably her favorite 3-D shape, much as the circle is in 2-D.

You're not sure as yet why she's so fascinated; left to herself, you suspect she might populate the universe entirely with circles and spheres.

But you're not going to leave her to her own devices. You have a thesis to prove, a universe to build. It's a daunting task: all the designing that lies ahead of you, using only mathematics. You feel you've made a promising start—laying down space, figuring out a language to orchestrate what you'll construct in it. Nature seems to be coming along quite well—she's no longer the novice mathematician she was when she started. You can see her developing into an effective contractor, one who will translate all your instructions into physical reality.

Unfortunately, things rarely go smoothly when a contractor is involved. Just ask anyone who's ever tried having a house built!

Day 4

PATTERNS

Formulating the underlying patterns of the universe

20.

PATTERNS AND PERFECTION

IT'S BEEN THREE DAYS, AND YOU'RE RIDING HIGH. You have a contractor all set to physically realize the universe you're designing, and the freshly minted language of algebra to specify and help build geometric shapes. Your first order of business for Day 4 is to assess how smoothly your work with Nature will proceed. Beyond that, it's time to brainstorm what else she'll need in the constructions to come. Mathematics, as mentioned earlier, is often called the study of patterns. What deep-seated patterns should you weave into the fabric of your universe? What are their properties, their aesthetics, their uses? How do you help Nature generate them?

Contractor blues

First, let's check on how well Nature's coming along as contractor. Alas, troubles have already begun to emerge. It's one thing to draw perfect shapes by activating coordinates, quite another to actualize these in your universe.

Take straight lines. Nature's great at calling out their equations to light them up in your coordinate system. But that's something you could do yourself just fine. The reason you hired her was to translate such lines into physical reality—for instance, by arranging a row of simulated particles along the coordinates. Such trials, unfortunately, do not end well. For one, the particles can't ever seem to keep completely still—their movements, though tiny, keep throwing the geometry off. Worse, Nature has only a finite number of particles at her disposal, so instead of being unbroken, the line is all fits and starts. It's also too thick compared to one made of idealized points, it's finite instead of endless, and so on.

You realize such limitations may be unavoidable in the translation from mathematical theory to a physical universe. But is Nature entirely free of blame? She shrugs in helplessness about the particles, but why *can't* she clamp down on their movement? Or simulate more of them? Some of the particles deviate so much from the line that she must have surely positioned them there on purpose.

Such aberrations get even more apparent in other constructions. Nature's squares have fuzzy boundaries, her spheres are dimpled and pocked. You increasingly suspect she might be adding her own alterations and embellishments. Is it to ward off bad luck, like a superstitious weaver incorporating intentional flaws in a carpet? Or just to assert her independence? To refuse to be bound purely by your mathematics, imprint her own sensibility on your universe?

You bristle. It's *your* design, not Nature's, that should shape the universe. But how to align her will with yours? You try to sell her on the beauty of mathematical precision—to which she just nods and continues as before. Demands that she'd better follow directions barely register, and neither does trying to sweet-talk her. There's no money involved that you can threaten to withhold. And forget about firing her—she knows she's the only outfit in town.

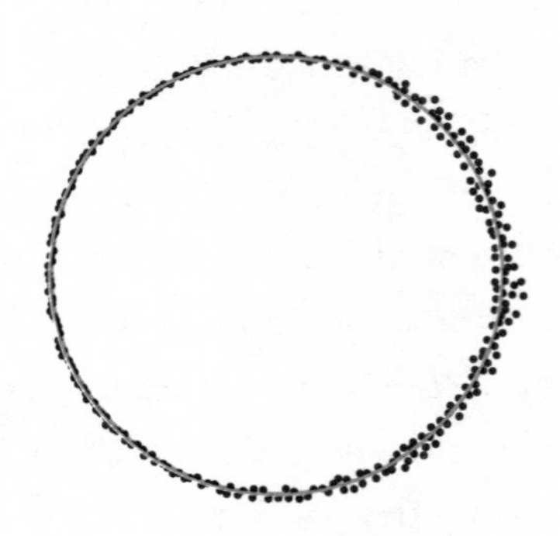

As she determinedly continues to turn out inexact samples with her particles, you get increasingly discouraged. The whole idea of this exercise was to create a universe built exactly in math's image, one embodying its perfection. A circle, for instance, represents an aesthetic icon—how frustrating to not be able to realize it. Perhaps this project was ill-thought-out, depending as much as it does on an intermediary with her own agenda. If Nature's going to continue with her impreciseness, perhaps you should just pull the plug.

Just as you're thinking of abandoning your universe, the pope intervenes to save it.

The papal rescue

Had you apprised him of your plans, the pope might have been able to warn you about what to expect. After all, he's familiar with the long history of problems his predecessors had with contractors hired to build the Vatican. All the prima donnas who worked on St. Peter's and its environs—Bramante to Raphael to Michelangelo—the latter even declaring he'd refuse to take on the project unless guaranteed absolute control. And then there was Bernini, the king of over-the-top, whose bell towers had to be demolished, whose extravagance almost emptied out the papal coffers. The pope is relieved not to have to fight such battles, nor bring artistic egos under control.

And yet, look how things turned out. The magnificence of the Vatican, its recognition as an artistic treasure, could hardly have been achieved by papal dictum alone. True, Pope Julius II set the ball rolling in 1506 with his vision of erecting the grandest holy monument ever. But it was the artists and architects and craftsmen who gave the project its beauty, brought it to resplendent fruition. Contractors are essential to creation, the pope knows—true glory can't be achieved without their interpretation, their ego.

In your case, it's not just any contractor but Nature herself who's going to give form to your design. Whether you think of her as God or science (the pope's choice is obvious, but he's at peace with dissension), the fact is, she's going to be powerful. So what if she operates with a touch of whimsy, if she doesn't exactly follow your mathematical dictates? The universe will be more interesting if infused with her caprice.

The pope thinks back to all those years ago when he was studying chemistry. He delighted in learning about the different ways compounds reacted, how every interaction could be represented by an equation involving symbols for individual elements. Although each equation was true, he realized, even way back then, that their aggregate could never describe all of reality. Nature had incorporated too many secrets into the mix, secrets she revealed with great stinginess. Whether it was because the magic of divine creation was unfathomable, or because scientific models could approach the truth but never quite clinch it, humans were destined to always remain unsated as they tried to penetrate Nature's mysteries.

Mathematics, in this respect, is similar to chemistry. While x might equal $(-b \pm \sqrt{b^2 - 4ac})/2a$ (the pope is proud, after all these years, to recall the quadratic formula exactly—though he did take the precaution of checking his answer on the internet), not every process in the universe is expressible so succinctly. There's a chance, of course, that everything *can* be reduced to formulas, that he'll be proved wrong—the Galileos of the past have taught him to approach his beliefs with humility. But in his heart he's certain this won't be the case, that his youthful realization was an immutable truth of the existence he and the rest of humanity share.

Mathematics, though, is also different from chemistry, harboring far deeper conundrums, as the pope realized back then. He remembers all the time he'd spend contemplating something as commonplace as a triangle. While he could conceive of a perfect triangle in his mind, even give a precise mathematical definition, such an object could, paradoxically, never exist in the physical universe. Every actual example—whether etched in pencil or with diamond tip, whether appearing in a crystal, an architectural facet, or a biological template, whether at the subatomic level or cosmological in scale—was an approximation, never living up to the flawlessness of pure geometry. No natural process—in physics, chemistry, biology, geology—could ever construct a perfect triangle.

God, presumably, could. But He chose not to—a source of puzzlement. Instead, He implanted the concept in humans, giving them the power of abstraction—so that they could look at roughly triangular shapes and conjure up a perfect version in their minds. And yet, they could never physically express such perfection.

What would happen if humans became extinct? Would exact triangles also cease to exist if nobody was left to envision them? The pope recognizes this as another one of God's riddles. It's always tempting to slide into such mindbenders, lose oneself in competing arguments that have no resolution. He wishes he still had the time for such pleasures, as he did when he was a student. But there are too many pressing issues on his plate—rampant hunger, rising oceans, geopolitical turmoil—not to mention Vatican intrigues. He barely has the few moments he needs to beam out his message.

Don't be discouraged if Nature's circles aren't exact, he tries to convey. *I understand your predicament, I've been there myself.* Rejoice in the idea of perfection, but also in Nature's approximations. Be open-minded to her rendition: it will create a richer universe. Besides, there's no other choice.

Thanks to the special telepathic bond I established with the pope during our *NYT* "top ten" buddy days, his words come in loud and clear. We'll follow his advice. He's the pope, after all, and if you recall, I said in the introduction that we want tips on how to make math as popular as religion. So from now on, let's avoid friction with Nature. Let's stop carping about her inexact geometry, or the number of her particles being finite. Perhaps, left to herself, she'll decide her universe really ought to be infinite. Our aim right now is to get to know our contractor, figure out her tastes and affinities. We want to find common ground on which we can collaborate.

21.

NATURE'S SOFT SPOT

SO, WHAT, IF ANYTHING, HAVE WE OBSERVED SO FAR about Nature's likes? (Other than her penchant for aberrations, which will come up again a little later.) Heading our inventory is surely her fondness for circles and spheres. This is what she lit up the most of when you told her to experiment on her own to get the hang of coordinates. Could it be the curvature of such figures that attracts her?

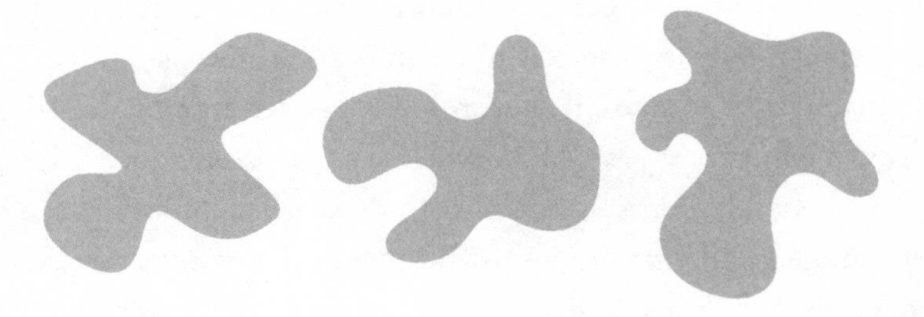

As a test, you sketch some extravagantly curved doodles. She barely gives them a glance. So it couldn't be just the simple fact of having curvature that she finds appealing. There must be something special about circles that turns her on.

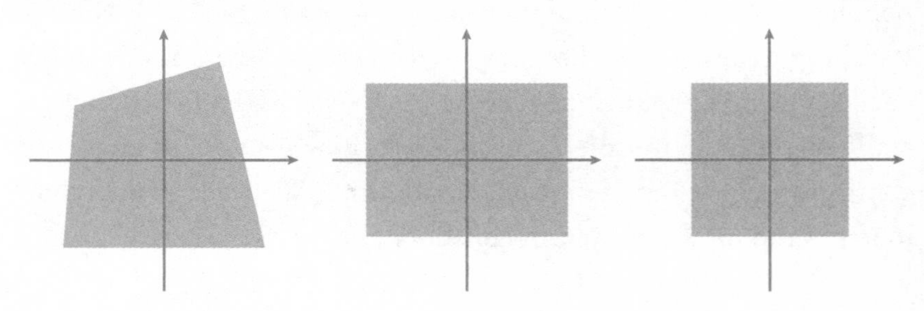

You've also noticed some definite preferences in the straight-sided figures she's lit up. Had she been drawing quadrilaterals completely at random, they should have almost all had four unequal sides each. But you see many more rectangles, and particularly more squares, than pure chance would create. She similarly reveals she's more partial to triangles with two sides equal—and even more so, three. Is there some common explanation for all these curved and straight-sided affinities?

You realize what Nature is expressing is a liking for *symmetry*. This is a concept we're all intuitively familiar with: each time we look in a mirror, we notice it in the left and right halves of our face. But how does it arise in the context of the universe-from-scratch we're building? How do we describe and quantify it? Why would our contractor be attracted to it?

Earlier, we saw how irrational numbers were responsible for introducing the first ideas of randomness into our universe. Let's see how, in a similar vein, the mathematics we've been developing gives rise to the idea of symmetry.

The math of symmetry

The earliest hints of symmetry were already surfacing when we had only numbers—for instance, in the way negative integers could be symmetrically paired with the positives, or the fact that order didn't matter in addition or multiplication. The latter property gave equations like $1 + 2 = 2 + 1$ and $1 \times 2 = 2 \times 1$, which, like palindromes, are symmetric about their center, the "equals" sign.

But symmetry really came into its own once geometry was introduced. Now the mirroring between the positives and negatives could be clearly seen when the numbers were arranged in a straight line about zero.

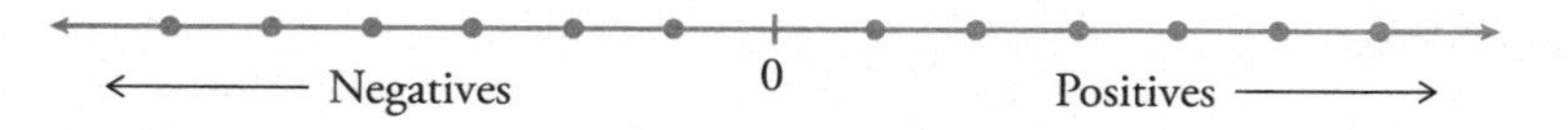

A way to characterize this symmetry is to note that the condo units for the positives are indistinguishable from the condo units for the negatives—so if flipped (or reflected) about the point 0, the line appears exactly the same.

This kind of mirror symmetry can be extended to figures on the plane. We can measure their symmetry, and even compare it, by asking the question: How many axes does the figure have, such that it remains unchanged if flipped about them?

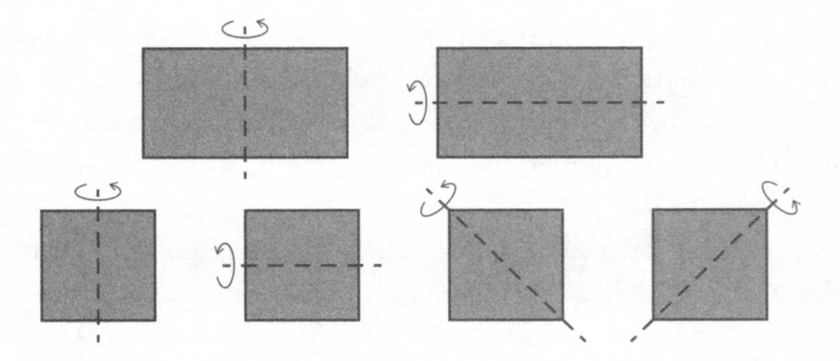

Rectangles, you will see, can be flipped about two lines of symmetry (the vertical and horizontal midlines). But squares best them by having *four* such axes, since the diagonals also do the job. So the square is more symmetric, which could explain Nature's attraction to it.

Each figure is also symmetric in another way—it can be rotated around its center to end up occupying the same place. The rectangle has two different rotations: a half and a full turn; the square has two additional ones: a quarter and a three-quarter turn. Once more, the square wins.

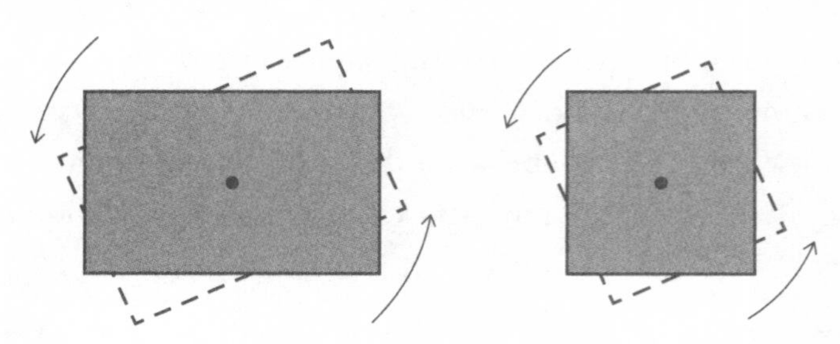

Other figures display such symmetries as well. For instance, an equilateral triangle has three of the mirror type and three rotational; a regular (i.e., equal-sided) hexagon has six of each kind; and a regular hundred-sided polygon (called a hectogon) has a hundred of each kind!

But Nature is unlikely to be churning out hectogons anytime soon. That's because if she wants lots of symmetry, she can just go for a circle. *Any* line through the center is then an axis of mirror symmetry, and *any* rotation, no matter how tiny or big, leaves the circle unchanged. In other words, the circle has an infinite number of symmetries; plus, it's so much easier to build. So if Nature's a symmetry nut, she's going to be particularly enamored of this shape!*

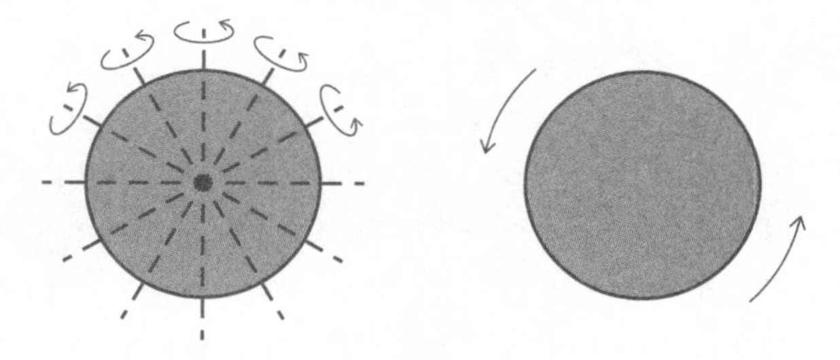

Could there be something even more symmetric than the circle? Well, you could go to 3-D, where the circle's analog, the sphere, can be mapped onto itself across an infinity of planes (titillating Nature even more). But what if you remained in 2-D?

There *is* something more symmetric than a circle in 2-D, and that's the entire plane. Suppose you wipe every figure, every mark, every number clean from it. You'll end up with a blank slate extending endlessly in every direction. Then reflection across *any* line or rotation about *any* point will just give back the same plane. (For the circle, the reflection has to be across a line through the circle's center, while the rotation has to be about the center itself.)

---

* Note that due to the inexactness she slips into anything she builds, a hectogon, for Nature, might be indistinguishable from a circle anyway!

But there's more. Since the entire plane is devoid of any features and infinite, to boot, if you shift the entire expanse in any direction, it will look as if nothing has changed. This is similar to what you can do with an endlessly repeating pattern—the infinite checkerboard below remains unchanged if displaced a whole number of squares in any direction. For an empty plane, the pattern is blank, so *any* amount of shift works. Therefore, in addition to symmetry when mirrored or rotated, the plane also remains the same when *translated*.

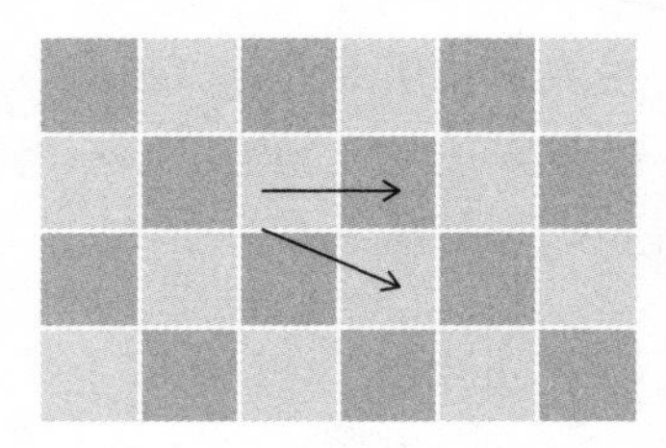

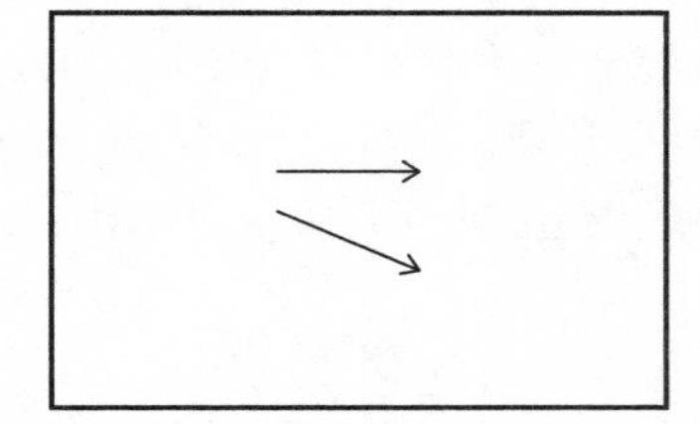

Does this blank and completely symmetric state bring up a memory? Perhaps of the nothingness we started with? The nothingness that you corralled to create Zero, then One, then everything. The universe of numbers—and all that followed—emerged from your act of breaking symmetry.

The lazy contractor

OK, we've noticed Nature likes symmetry, but why? One possible reason is she's lazy,* and symmetry cuts down her work. With a mirror-symmetric figure, for instance, she can generate just one half, then reflect it to get the other. This might be one of the reasons our

* More generously, Nature (as physics or God) is frugal in terms of expending effort. Certainly, physics is driven by all kinds of conservation (e.g., of mass and energy). Deities, too, feel the exertion. As noted earlier, Genesis has God resting on Day 7, while in Hinduism, Vishnu lapses into deep slumber in between the universe's successive life cycles.

own bodies are symmetric: If you have a working blueprint, why not replicate it?

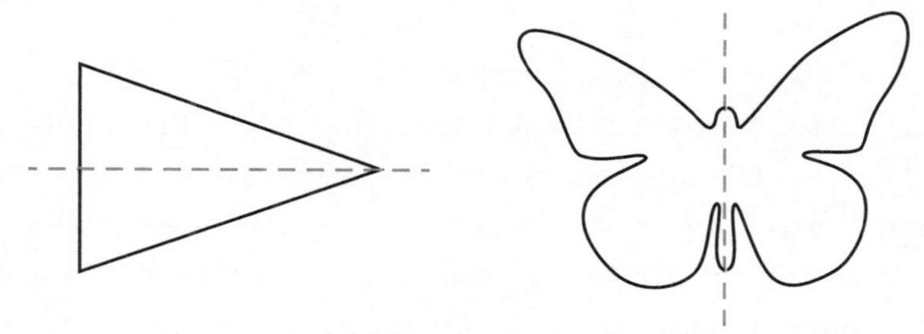

With more axes of symmetry, reflections can be repeated, making Nature's workload even lighter. For instance, for a rectangle, she could construct just a quarter of the figure, reflect it across a vertical line to copy it, and then reflect the aggregate across a horizontal line for the full shape.

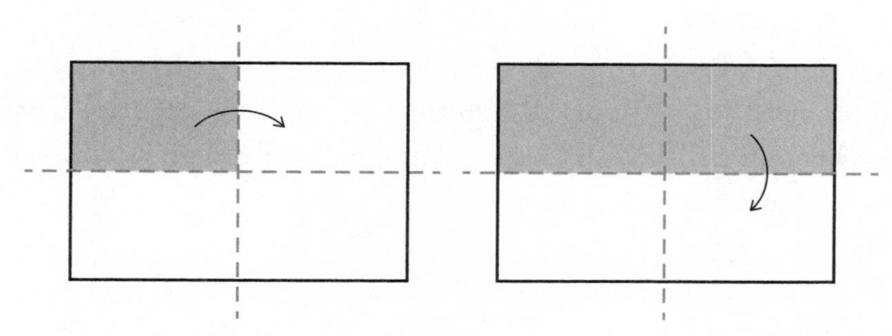

Carrying this idea further, a hundred-sided hectogon could be constructed by drawing just a tiny sliver, then reflecting it many times. But this would surely be overkill (and tedious, as well). So while Nature might find a few mirror axes useful, this doesn't tell us why she would be enamored of the abundance of symmetry displayed by a circle or sphere.

Here's a possible explanation. Imagine, first, Nature's reaction to the most symmetric thing possible in 3-D—the entire, endless, empty space. No matter where she is in this space, it looks the same in all directions. Her view does not change if she walks about or turns around or performs cartwheels. The perfect symmetry of emptiness makes it very easy to interact with—whether Nature's moving in it, building in it, formulating laws in it, or performing any of the tasks she expects you'll soon ask of her. There's never a change of orientation to process.

Now suppose a sphere is introduced into this space. Then, unlike before, Nature's view will change depending on her line of sight. As an object, though, the sphere is very simple for her to process—it has an inside, an outside, and a surface that is exactly the same all over. This sameness makes it easy not only to recognize but also to incorporate into any project—she doesn't have to worry about any special feature or anomaly. Contrast this with a cube, where she'd have to keep track of a much more complicated surface involving faces, edges, vertices, and various junctures among these. She'd need some mental reconciliation just to recognize it as the same object when viewed from different angles. Building up structures or laws involving the cube would be more onerous, since she'd have to worry about how things might change near its various features—perhaps the faces would need to be treated differently, perhaps some singular effects would occur at the edges. Fortunately, the six-sided symmetry would reduce the bookkeeping somewhat.

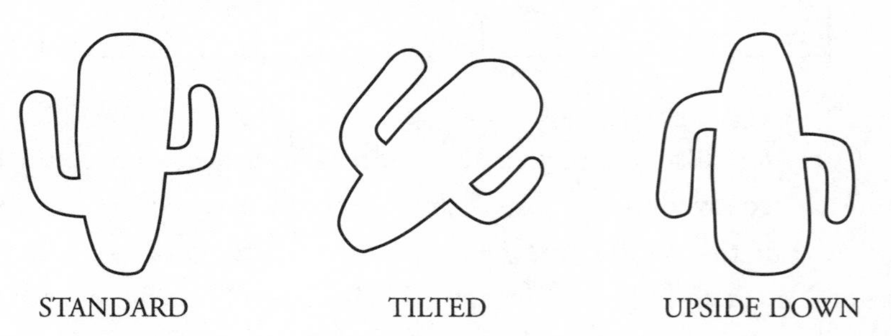

If the cube is replaced by an arbitrary object devoid of symmetry, then Nature doesn't get any breaks. For instance, consider the asymmetric 2-D cactus shape shown. Depending on how Nature orients herself, the image will look tilted or even upside down. To recognize it, she has to compare it with the original—say, by designating a top and bottom, a right and left, and then matching each part. As a contractor, she'd have to track not only all these regions (each of which might call for a different treatment in a project) but also detailed specs like the size of the base and the crook in each arm. Contrast this with the case of a circle, where no such complications arise—the top, bottom, left, and right are all the same.

Like any other contractor, Nature wants to minimize her effort. She'd rather not be inconvenienced by *any* specs. However, this works only with the pure symmetry of emptiness. Nature's not going to have the luxury of this for long; she knows her job is to create and work with objects. Symmetry, she feels, can help whittle down her work.

22.

THE GOLDEN RATIO

JUST WHEN YOU THINK YOU'VE GOT HER ALL FIGURED out, you realize Nature might be driven by more than just a love of symmetry. You've been observing her experiment some more with the coordinate system—light up shapes and figures as you encouraged her to, overlay these with strings of particles on occasion. Compared to the number of squares, she seems to be drawing more and more rectangles of late. Not just any rectangles—what surprises you is that although they come in different sizes and orientations, they all seem to have the same proportion!

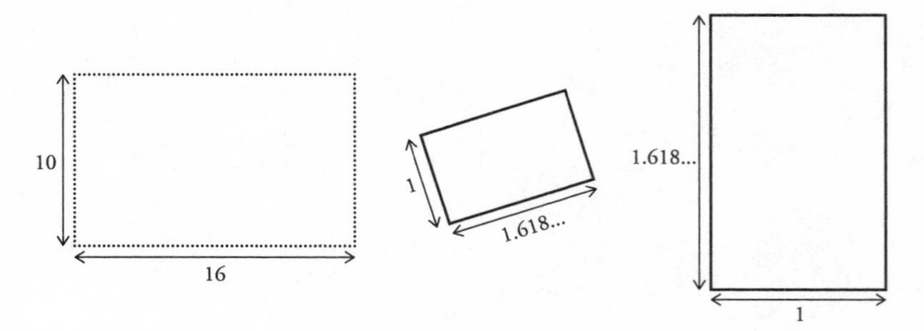

What is this proportion? At first, you notice that each time Nature arranges particles along the outline, she always uses about 16 particles along the longer side for every 10 along the shorter. Which means the ratio of lengths is around 1.6 to 1. Later, you're able to figure out a more precise estimate for this ratio—it's approximately 1.618. You're puzzled by where Nature might have come up with this strange number, and why she's suddenly attracted to it. The only other affinity she's shown is for the number pi, which she's discovered gives the ratio of circumference to diameter for her beloved circles. But 1.618? She didn't get it from you. Could she have learned it from the numbers?

The mysterious factor

You realize you've seen this ratio before! It arose long ago, from a game the natural numbers invented when addition first became popular. Two numbers would kick things off by adding themselves together. Their sum would then add itself to the number that came before, and so on. Repeating such successive additions, an unending train of numbers would emerge.

For instance, starting with the numbers Zero and One led first to $0 + 1 = 1$, which then gave $1 + 2 = 3$, $2 + 3 = 5$, $3 + 5 = 8$, and so on. This led to a sequence in which each number was the sum of the previous two:

$$0, 1, 1, 2, 3, 5, 8, 13, 21, 34, 55, 89, 144, 233, 377, \ldots.$$

Starting with different initial pairs led to different sequences—for example, Two and Nine resulted in

$$2, 9, 11, 20, 31, 51, 82, 133, 215, 348, 563, 911, 1474, \ldots.$$

You recall the game became quite popular, since it provided a fun way to get familiar with the newfangled addition. The lists generated changed so much with different starting values that it became a guessing game to see whether some larger natural would make an appearance in a list or not.

With the invention of division, a surprising common pattern behind these diverse lists was discovered. When neighboring numbers far enough down the list tried finding their quotient (with the bigger number on top), it always turned out to be about the same. For instance, taking successive pairs with the last few entries shown in the first sequence above gives (roughly)

$$\frac{144}{89} = 1.6179, \frac{233}{144} = 1.6181, \frac{377}{233} = 1.6180.$$

while with the second, we get

$$\frac{563}{348} = 1.6178, \frac{911}{563} = 1.6181, \frac{1474}{911} = 1.6180.$$

Over and over, no matter which two naturals started the game, the numbers saw the same quotient of about 1.618 emerge. This was none other than the ratio Nature had chosen to now draw her rectangles with!

Surely this couldn't be a coincidence. The naturals must have supplied Nature with this ratio. But why had she adopted it? The naturals played a bunch of other games, and some of these led to other ratios.* What had attracted Nature to this particular ratio? What was special about it?

Self-symmetry

The answer comes one day when you see Nature have the numbers light up a curious formation with their points. The rectangle with the 1.618 ratio is still there, but now there's an extra line added to mark off a square from it.

---

* For instance, in a variation, the numbers started by adding *three* initial naturals together, and creating a sequence where each entry was the sum of the previous *three* entries. This yielded a corresponding ratio of about 1.839. You can check this out yourself.

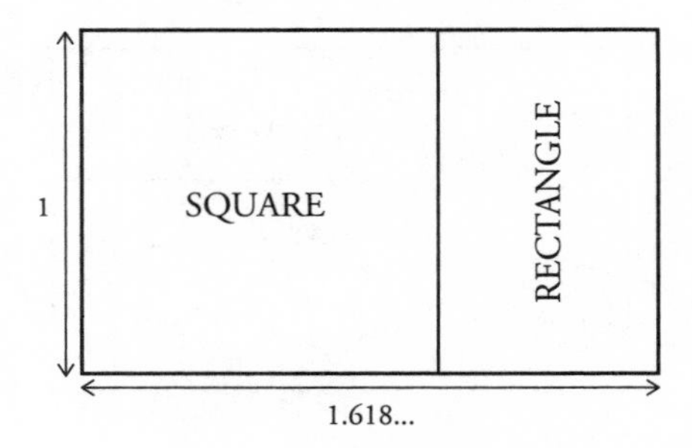

This leaves a leftover piece that is rectangular. When you look at this piece more closely, you realize something very interesting: this leftover rectangle has the same 1.618 ratio shape as the original rectangle, except it is smaller, and rotated a quarter turn. The more you think about this, the more remarkable it seems. Subtract a square from the original shape and you get the same shape back!

Surely rectangles with other proportions might also display this property? You try drawing a few, but subtracting a square never gives a rectangle similar to the original. You're able to convince yourself that 1.618 (or a number very close to it) is the only ratio that works. In fact, this process of subtraction can be repeated to give a series of smaller and similarly shaped rectangles that fit within each other like Russian Matryoshka dolls. Nature seems to have chosen the mother of all rectangles!

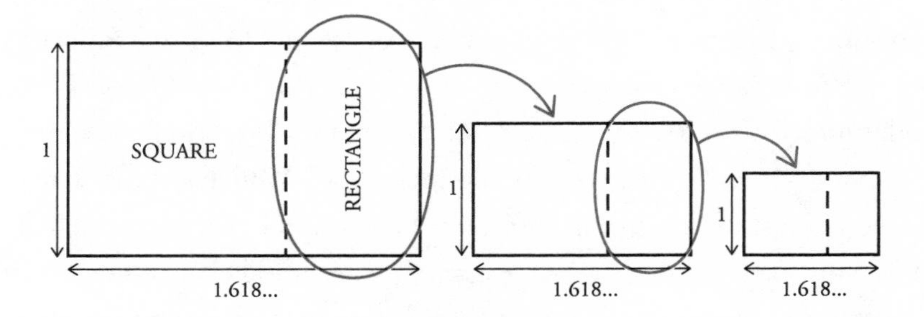

For a while you think you've solved the mystery—that Nature's been enchanted by the self-symmetry of these nesting Matryoshka figures, and that's why she's chosen this rectangle. But the reason is deeper, as you find out one day when Nature lights up a chain of nested rectangles.

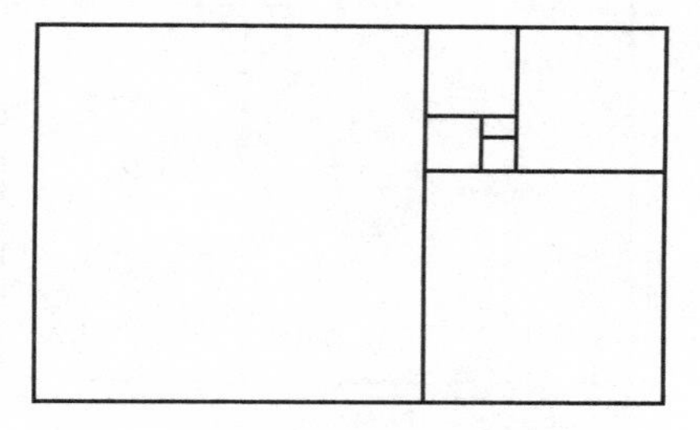

As she's adding lines, you realize she could keep going: the outer rectangle can be divided into an infinite number of squares of decreasing size (try adding the next line or two yourself in the above diagram). Nature, however, sticks to the finite, stopping once the divisions become too small to detect. She then switches to her particles, and starts arranging them in the form of a curve guided by the squares she's lit up. This curve is not a circle, but something new you've never seen before in your universe—a *spiral.*

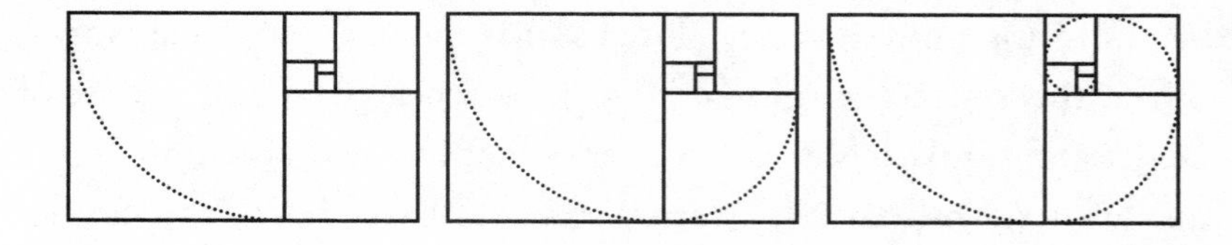

There's something mesmerizing about this spiral—the more you look at it, the more you're drawn in, as if watching a hypnotist's wheel. You realize that were you to zoom in, the curve (if continued indefinitely) would keep repeating the same shape, at smaller scales—it inherits the self-symmetry of the rectangles. Which means that to draw it, Nature doesn't have to alter her design—the same instructions continue to work for every stage. Is this why she's attracted to it? Another manifestation of her laziness? Or, to be more polite, frugality of effort?

Laziness or not, what's astonishing is the fact that Nature has figured out how to create such a figure. Just yesterday, it seems, you were trying to teach her to draw a straight line. It's remarkable to see how adept she's become in putting mathematical ideas to new uses, how fast she's progressed from awkward student to inventor.

Which makes you feel a little insecure. What if she stops needing you? What if she goes rogue and starts doing her own thing entirely? She's always hinted she wants to be credited for creating the universe. What if she stops being your contractor?

23.

NATURE'S RELATION TO MATH

We're going to pause our thought experiment here for a discussion break. That's because the ideas introduced in this chapter are some of the most publicized in popular mathematics. The number 1.618 . . . is the so-called Golden Ratio or Divine Proportion, the favorite poster child of the link between math and Nature, supposedly present in everything from gender ratios in beehives to the shape of ancient pyramids. In fact, there's even been a best-selling potboiler (*The Da Vinci Code*, with its obligatory Hollywood film adaptation) larded with such claims. This is the perfect spot to stop and ponder the true relation, get a better idea of what math really is.

The Fibonacci rabbit parade

To do so, let's go back to the sequence 0, 1, 1, 2, 3, 5, 8, 13, . . . , which gives approximations to the Golden Ratio when you divide successive terms. This sequence is named after the Italian mathematician Leonardo Fibonacci, who introduced it in his 1202 book, *Liber abaci*,* in the context of trying to figure out how rabbits multiply.

---

* Fibonacci's greatest impact on mathematics was not the sequence he's remembered by, but the fact that *Liber abaci* (Book of the abacus) introduced the Hindu-Arab numerals to many in Europe, and helped supplant the unwieldy Roman system in use before. (Try adding LXXXIX and XCIV, for instance. Or worse, multiplying them.)

Assuming rabbits came in male-female pairs, he asked how many such pairs there would be in a year if each pair took a month after birth to reach reproductive maturity, and then gave birth to a new pair each month after that.

Let's unpack this rule. Suppose we begin with one newborn pair in month 1 (in month 0 before, the population is 0). Month 2 is taken up with this pair maturing, so the population remains 1 (the sequence of monthly populations is therefore 0, 1, 1 so far). In month 3, this pair, now mature, gives birth to a new pair of rabbits (the sequence becomes 0, 1, 1, 2). In month 4, the newborn pair matures and the original pair gives birth to another pair of rabbits (0, 1, 1, 2, 3). Month 5 is interesting because both the original pair and the second pair (now sexually mature) give birth, so there are *two* extra new pairs born (0, 1, 1, 2, 3, 5). Continuing this way, month 6 will have 3 new pairs born (0, 1, 1, 2, 3, 5, 8), and so on. We get the Fibonacci sequence.

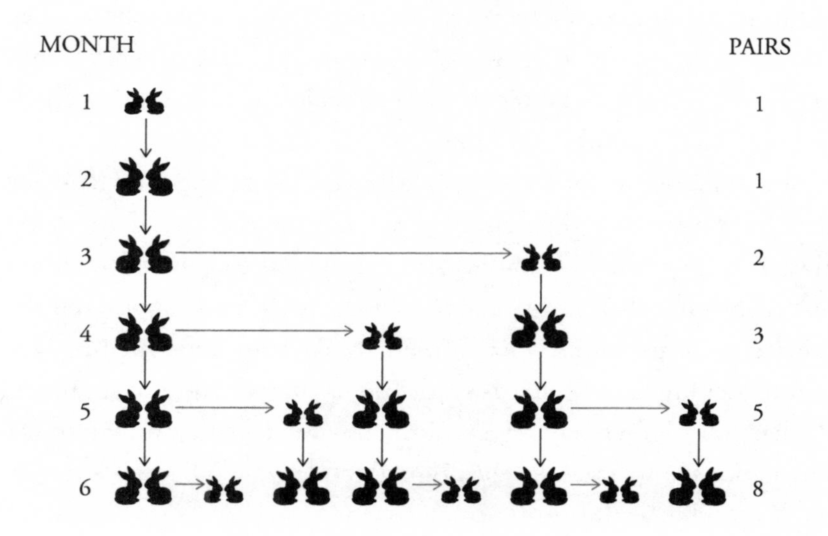

But where exactly does the Golden Ratio manifest itself in all this rabbit randiness? To answer this, recall that if one goes far enough along the Fibonacci sequence, then the ratio of successive terms is about 1.618. This means that if x is some term in the sequence, then the next term will be approximately $1.618x$. So the same must be true for the rabbit populations after a while, that is, they will settle down into growing more or less according to the rule

$$x \rightarrow 1.618x$$

each month.* Indeed, the predicted populations for months 12 and 13 are 144 and 233 respectively, and 233 is just about 1.618×144.

Does this work in real life? If you start with an actual pair of rabbits in month 1, will you get 144 pairs in month 12, as Fibonacci figured out from his sequence? Will the population start increasing by a factor of 1.618 or so each month?

It won't. For one, rabbits aren't born neatly in male-female pairs. For another, they won't follow the monthly birthing rule with such machine-like precision. Different litters may be larger or smaller, pregnancies may occur later or earlier. Then there's the whole issue of mortality—unless we have a special breed of vampire rabbits, they might not be around long enough to make the numbers work. Finding perfect Fibonacci rabbits is as impossible a task as finding a perfect physically manifested triangle.

What Fibonacci provided was a *model* for rabbit growth—one that, like a perfect geometric figure, is a mathematical ideal. As always, reality is more complicated. Nature has to contend with not just rabbits but their individual variations, their food supply, and the prevalence of predators who eat them. She has to fine-tune in a web of interlocking factors such as space and habitat, the heat of summers, the harshness of winters. By the time she's woven everything into her calculations, only traces of the original model would be left—the formula would have drastically changed.

---

* This is an example of an *input/output rule*, where the input of x rabbits at the start of the month ends in an output of $1.618x$ rabbits at the end of the month. We'll be seeing several more rules in this format ahead.

Exponential growth and the pandemic

And yet, Fibonacci's model contains an essential truth about how populations grow. The rule $x \to 1.618x$ can be written more generally as

$$x \to (\text{growth factor})x$$

where the growth factor can be any other number, not just the Golden Ratio. This rule is the defining equation for so-called *exponential* growth, observed in many natural phenomena (Fibonacci growth is a special case). For instance, the growth factor is often taken to be 2 for bacteria growing in a petri dish or viruses infecting a human population, since both of these, at least in the early stages, are known to double every time period (this period's length being determined by observation).* The same rule also tells you how the "population" of dollars in your bank account grows when you get interest. For instance, the growth factor would be 1.07 if you received 7 percent interest compounded every year.†

How well do such rules predict nature in practice? The caveats with Fibonacci's rabbits might make you think not very well, but as we have seen with the COVID-19 pandemic, such growth models are indispensable tools, and can be very accurate when properly calibrated. Look at how the number of COVID infections rose in the early days in almost any country affected, and you will see close adherence to exponential growth. The tally for the entire globe was

---

* Actually, *any* exponential growth can be modeled with a growth factor of 2. What this means is that the population will double over each of successive time periods, you just have to find the right period. For instance, rabbit populations following the Fibonacci sequence double every month and a half (more precisely, every 1.44 months) if the growth is represented as a continuous graph. (For why it's called *exponential* growth, see the endnotes.)

† Calculations show that with this interest rate, your money would double about every 10 years (10.24 years, to be more precise).

also modeled very accurately by exponential growth, as the graph below shows.

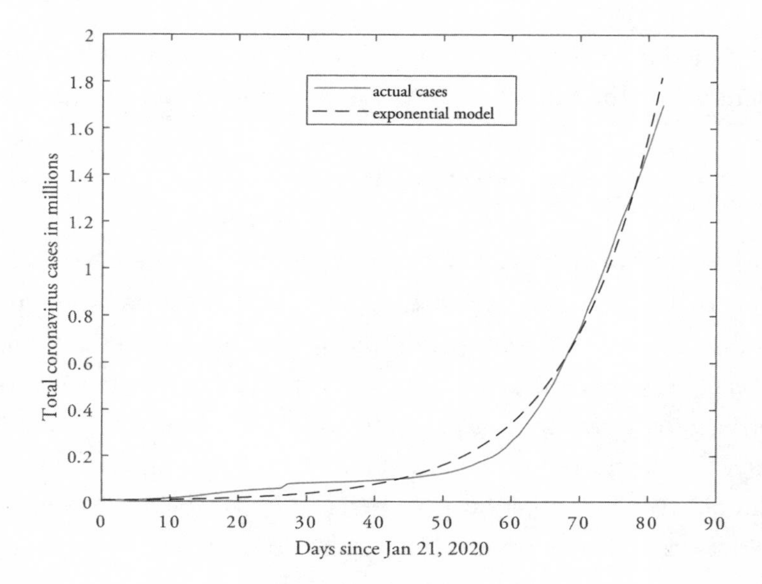

Such growth stems from an elementary mechanism: each infected person infects, on average, a fixed number of others in a given time period. Any increase satisfying this basic assumption (implicit in the "$x \longrightarrow$ (growth factor)x" rule) will *always* be exponential. This rocket-like rate abates only when the growth factor decreases—that is, the number of new infections per infected person diminishes (due to social distancing, vaccinations, immunity, and so on). In fact, tracking this rate for reductions in the growth factor (characterized by a flattening of the curve) yields an indication of how well preventive measures might be working.

The pope will never forget the Easter sermon he delivered in 2020. It was the first Easter during which the pews of St. Peter's Basilica stretched out almost empty before him. His homily conveyed the hope of resurrection that the day symbolized, the conviction that the world would recover and heal. While these words of faith were live-streamed to millions, there was a parallel message being conveyed by mathematics. New infections in Italy had recently turned a corner, and calculations showed the growth factor was declining. An idea proposed to predict rabbit populations was being used to chart this decline. As it turned out, this abatement was temporary, and there were worse waves to come. But in all these ups and downs, mathematical analysis provided eloquent reassurance to the planet, in a very different way from the pope's, that the coronavirus could be checked.

So yes, Fibonacci uncovered something essential about how populations can be modeled in general. But let's get back to the Golden Ratio. Where, specifically, does it pop up in nature?

Spirals

Technically, the ratio 1.618 . . . can *never* pop up exactly in Nature's physical manifestations. That's because it's an irrational number (as indicated by the dots),* and hence requires an infinite number of digits to express it (much like pi, which we can also never expect to come across, exactly). But that's fine, because Nature, as we've seen, is pretty inexact anyway, tending to blur and smudge whatever comes her way. So the question should be modified to "Where does a *close approximation* of the Golden Ratio pop up in nature?"

If one just looks for Fibonacci numbers (rather than their ratio), then perhaps they appear most conspicuously in botanical spirals. For instance, the hexagonal segments on the surface of pineapples are arranged in spirals, as are the scales of a pine cone, and the seeds of a sunflower. Count the number of such spirals, and you'll find they're most often Fibonacci numbers. As an example, the pineapple shown has 5 spirals that are close to horizontal (these curl around the back, like number 5, which is shaded), 8 spirals that are more tilted, and 13 spirals with the greatest slant (counting around the pineapple). Simi-

---

* Euclid was the first to show its exact value is $(1 + \sqrt{5})/2$, which he obtained by algebraically solving the "subtract a square from a rectangle" problem (see endnotes for this solution).

larly, the pine cone shown is partitioned into a pattern that is seen to have 13 spirals.

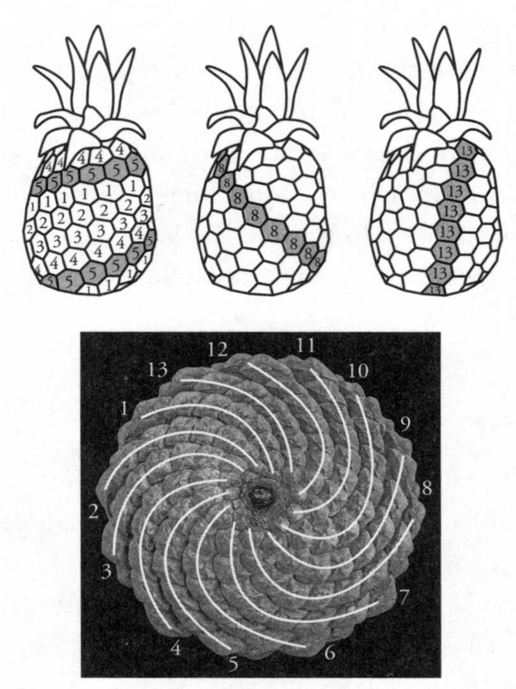

As for the ratio itself, a critical appearance it makes is in phyllotaxis, the arrangement of leaves around a stem in plants. It does so in the form of the Golden Angle, defined as the smaller angle formed at the center when you divide the circumference of a circle into two pieces in the ratio 1.618 . . . to 1, as shown in the diagram. This comes out to be about 137.5 degrees—the exact value is again irrational.

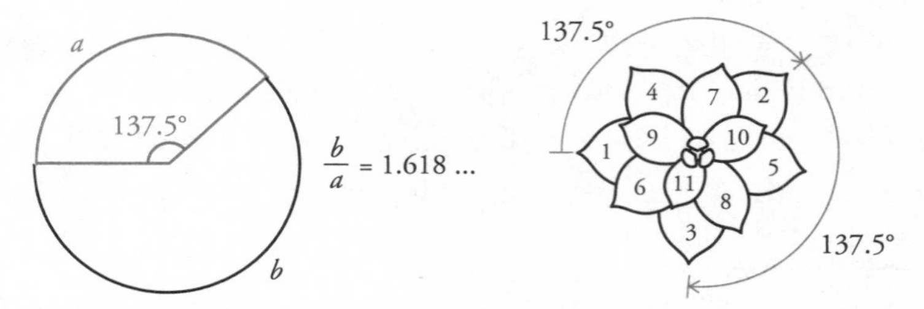

This angle is observed at the growth tip of many plants, separating new leaves from one another, as they emerge in a circular pattern. As the stem elongates, the leaves form a spiral whorl around the plant. (The diagram shows the view looking straight down—leaf 1 is at the bottom, leaf 11, the most recent to emerge, at the top.) The Golden Angle turns out to provide the most favorable spacing in terms of leaves higher up not shading the ones below, thus maximizing the opportunity for photosynthesis.*

Since we're talking about spirals, let's recall the special Golden Spiral constructed by Nature at the end of the last chapter. Can we find it physically manifested in our universe? Certainly, there are several claims saying that both the chambered nautilus shell and spiral galaxies follow this curve.

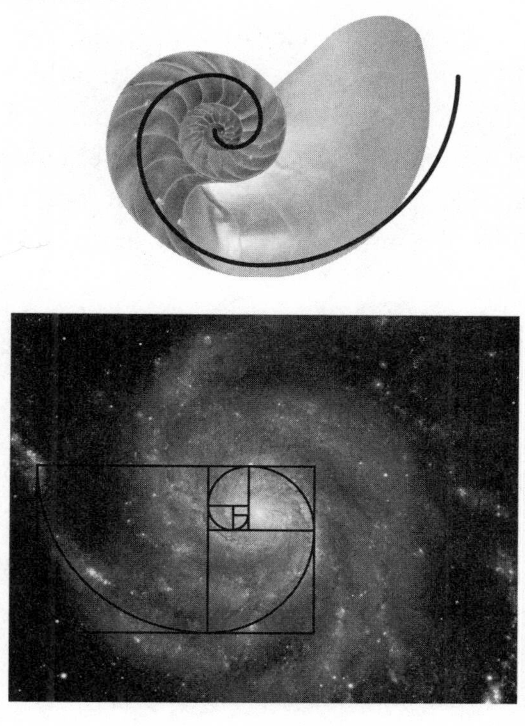

* For example, if the angle had been 90 degrees instead (which is 360/4), then the leaves would all be neatly stacked up, directly over each other, in four vertical columns. Similar vertical columns would be formed by angles like 360/2, 360/3, 360/5, etc. But 137.5 . . . results from dividing 360 by an irrational number, and is not close to any of these rational number quotients. Since leaves can no longer settle into a repeating pattern, this angle provides more spread-out spacing.

As seen earlier, the Golden Spiral has the special property of self-symmetry, which means it does not change its shape as it winds in or opens up. This is a very desirable property for a growing mollusk to adopt for its own shell: at each stage, it can now create a new, larger chamber to move into, and close off the older one, without having to constrain its own size in any way. The resulting nautilus spiral is not the same as the Golden Spiral, but rather, inspired by it—it's an example of a so-called logarithmic spiral, of which the Golden Spiral is the prototype. So the Golden Spiral leads to a family of useful models that Nature can add to her bag of contractor's tricks.

Galaxies also often unfurl in logarithmic spiral shape, though the reasons are different. Some scientific papers claim that the tendency is for galaxies to actually replicate the Golden Spiral rather than only a related configuration. Whether exactly golden or not, Nature seems to love logarithmic spirals, deploying them in many of her other works: the horns of rams, the eyes of hurricanes, the path a raptor takes toward its prey.

What this tells us about math

You can find accounts of several more instances of the Golden Ratio manifesting itself in our universe, beyond the ones I've given. Some of these examples embody the actual ratio, others are only inspired by it, still others have dubious connections at best. The overall evidence suggests that the ratio does have a special presence in nature, even if it's not the key to the universe, as some claim.

Which brings us to a fundamental question: How to interpret such connections?

The pedagogical conceit in this book is that we develop mathematical rules and convey them to Nature, who then builds the universe using our specs. So, once the Golden Ratio is revealed to her through the Fibonacci sequence, she incorporates it in her templates. In fact, she must take a particular shine to it, given that she deploys it over and over again.

This premise that we teach Nature math and she uses it to create is, I hope, proving to be illuminating. I adopted it because of the ease with which it allows us to explore how numbers underlie our universe—the goal I laid out at the beginning of this narrative.

But let me emphasize that this is very much a thought experiment we're in. Perhaps the universe could indeed be built step-by-step according to our mathematical construction, but that's not to say it actually happened that way. For instance, Nature falling so much in love with the Golden Ratio that she whimsically uses it everywhere is an appealing scenario, but the evidence for it is thin. Recall that the ratio crops up for specific reasons, and these can be different in different settings. In the rabbit example, it arises directly from the additive aspect of the Fibonacci sequence, while in the case of phyllotaxis for leaves, its appearance is related to the irrationality of the Golden Ratio. Nautilus shells mimic the Golden Spiral because of the shape and size advantages afforded by self-symmetry, but there's a separate explanation for spiral galaxies, and yet another for falcons spiraling toward their prey.

So, rather than regarding the Golden Ratio as being formulated abstractly through mathematics and then imposed on the physical universe, we can equally well (some might insist more plausibly) reverse this view. Mathematics is a language we develop to describe the universe. When we analyze the workings of the universe's various components, we detect various common rules and patterns that underlie diverse phenomena. These commonalities can be easily described using mathematics. The Golden Ratio emerges as one such artifact. The doubling factor, seen in growths we call exponential, emerges as another.

Perhaps the easiest way to summarize the two approaches is that in the first, mathematics is used to *create* the universe, while in the second, mathematics is used only to *describe* it.

Each approach presents its own answers and conundrums. The

first advances an answer to the question of why nature has rules and patterns at all. It explains the so-called "unreasonable effectiveness of mathematics" in describing the universe, as we mentioned in the introduction. However, it does not address the mystery of why Nature would choose to implement math so shoddily, often squandering its exactness. (Surely a child could have done a better job following equations?) Also, as we'll see on Day 7, it runs into problems explaining how plans could have been fine-tuned so precisely in advance to get the universe exactly right.

The second approach leads to great success at approximating the universe with an accuracy high enough to make predictions. The goal is not to capture the absolute reality of the universe (which is probably unachievable) but to be able to construct a sequence of models that can approach this reality with whatever level of precision is desired. Nature doesn't follow mathematics; rather, the subject is purely a human invention made to describe her rules. Mathematics allows us to feel we have some agency over her, whether it be to predict tides, or navigate according to the stars, or track pandemics.

Which of course leaves the "unreasonable effectiveness" question unanswered. Particularly mysterious are branches of mathematics that were developed entirely through abstract contemplation and later found to provide deep-rooted explanations of the universe. A primary example is the existence of different types of curved geometries that Einstein found so useful in modeling gravity and spacetime (we'll touch on these during Day 5). Another is the theory of symmetry (and related "group theory"), which turns out to play a fundamental role in the classification (even discovery!) of elementary particles from which all matter is built.

As if the above two philosophies weren't enough, we earlier mentioned a third way of looking at mathematics, subtly different from the first. Plato postulated the existence of a separate universe where all idealized mathematical concepts live. Not only will you find perfect triangles and circles there, but also every mathematical theorem, with elegant proofs to go along. Humans do not invent mathematics as we've been doing in our thought experiment to create the universe, or as scientists might when they construct models—rather, they only discover it. Mathematics has existed independently of us since time immemorial—and will continue to exist. Every once in a while, the clouds part to give us a glimpse of this universe, and that's how we learn about it.

So what is mathematics? You should be able to convince yourself that it incorporates aspects of all the views above but doesn't neatly fit into any one slot. We could spend many more pages pondering its exact nature, but instead, we'll stick to the premise we started with—that we formulate it, then supply it to Nature. As the pope realized long ago, there are too many intriguing mysteries in the world, ready to swallow up more time than we might have at hand to give.

24.

MATH AND BEAUTY

SINCE WE'RE PONDERING MATH'S INTRINSIC NATURE during this break, let's look at the following popular question: Can there be a mathematical criterion for beauty? This might seem unlikely, given that math is so objective and beauty so subjective. And yet the two most recent ideas we've looked at—symmetry and the Golden Ratio—have staunch proponents who advance them as ways to measure beauty mathematically. What we're going to do is pit these teams against each other in a competitive spectacle to see who wins. The pope's already ordered pizza for this halftime show—perhaps you can get some beer to go with it.

The Sym team touts symmetry. It points to experiments showing that people prefer human faces that are perfectly symmetrical. One theory for this preference is that we subconsciously equate symmetry with health, and asymmetry with illness (at the most extreme, an asymmetric face or silhouette might indicate a missing part). Another theory is that our brains are hardwired into processing symmetry more easily than asymmetry. (In this regard, recall that we discussed earlier why Nature might find it easier to recognize symmetric shapes over nonsymmetric ones.) Mathematically speaking, one can easily rate one object more symmetric than another by counting the number of mirror symmetries (for instance) in each. Does this give a measure of beauty as well? The Sym team declares yes.

The Ratio team, in contrast, plugs the Golden Ratio as *the* standard for mathematical beauty. It furnishes its own set of studies that claim that the ratio, not symmetry, is what humans find most appealing.* That's why the ratio is supposedly found in the Great Pyramid, the Parthenon, old master paintings, even the actress Angelina Jolie's face. Many of these connections are murky—for instance, the Great Pyramid and Parthenon yield the ratio only if one goes on a deep dive for it or massages measurements until they fit. However, it's also true that no less an artist than Leonardo da Vinci drew the sketches for a treatise on the subject. There have been several papers written on whether he used the Golden Ratio in the design of Mona Lisa's face.

Rather than wade through these papers, let's just assume he did, as evidenced by the boundary of Mona Lisa's visage fitting neatly into a Golden Rectangle. The Ratio team, which hypes this as the reason behind her appeal, cheers.

To be evenhanded, I now offer to help the Sym team. I suggest we try to prove that adding more symmetry to the painting would make Mona even more beautiful. My new teammates like the idea, even

* I'm exaggerating the conflict here. The Sym team's experiments with faces deal only with bilateral symmetry (i.e., on either side of a vertical centerline), and not other symmetries (e.g., across a horizontal centerline). The Ratio team's studies based on the Golden Ratio typically ask subjects to pick the most aesthetically pleasing figure out of a bunch of rectangles. The two teams' studies do not quite contradict each other. In the second type of study, rectangles tend to be more popular than squares, but whether the Golden Ratio is most popular is unclear. See Mario Livio's book *The Golden Ratio*.

though they don't know exactly what I mean. I explain that Mona's face as represented on the canvas isn't *exactly* symmetric. But it's easy enough to fix with cut and paste.

Are those gasps I hear? The Sym team is dismayed. All I did was reproduce the left part of the picture on the left to give it mirror symmetry. They should commend me, instead of voicing complaints. But they say the nose is too big now, the skull too pointed, the eyes unfocused, as if Mona's stoned. I assure them I'm on top of these problems, that all we need is a bit more symmetry. I perform my cut-and-paste magic again.

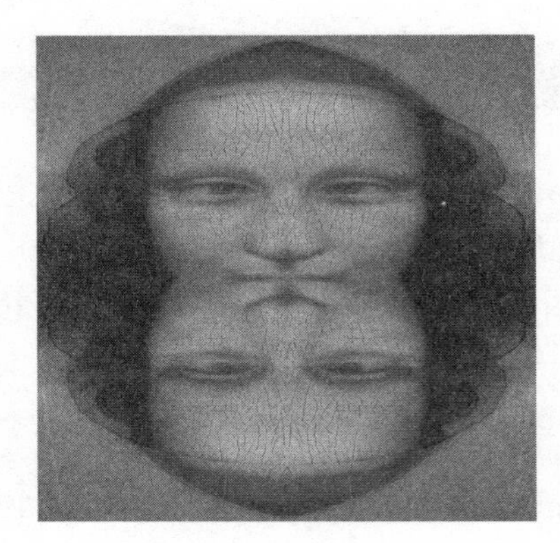

The team gets even more distraught. Despite my pointing out that we now have mirror symmetry across two axes! Isn't that what they said would lead to more beauty? I urge them to have more faith in their thesis. Don't chicken out—forge on, add even more symmetry. They watch in dread as I cut off a quarter of the picture, then paste four copies together for a new image.

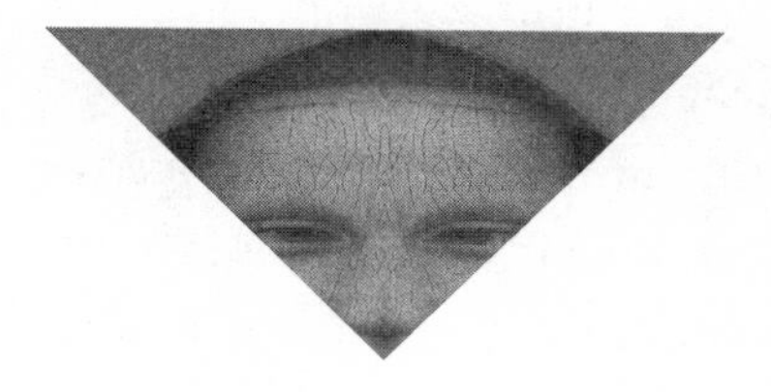

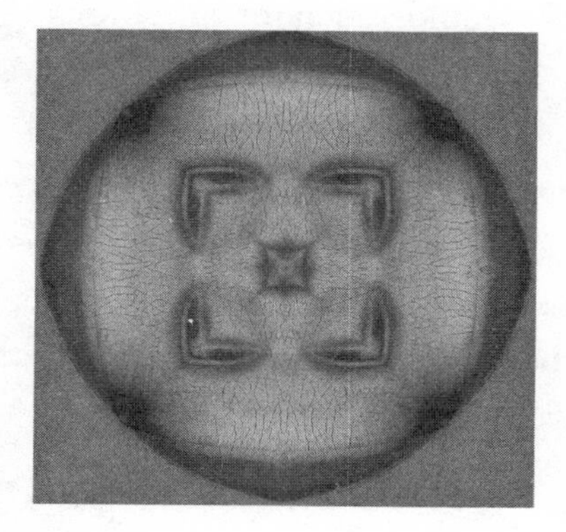

Mona's now symmetric across the diagonal as well—she and the square have the same number of symmetries! The transformation sends the Sym team into a tailspin. You've turned her into a bloated soccer ball, they accuse me. I try to lift their spirits—keep going, I urge—let's give her even more symmetry. I shrink the wedges, then paste eight of them together to form an octagon.

My new masterpiece gets little reaction—the Sym team has sunk into despair. I try to rally them some more: we're almost there, almost over the hump! I make the slices even thinner and paste them together. Mona Lisa is reborn as a twelve-sided dodecagon.

This time, some teammates seem to perk up a bit. Do you see it? I ask. Symmetry working its magic? It's a completely different form, but would you agree that Mona's getting beautiful again?

She reminds me of a flower, someone comments. My mother's tablecloth, another person says. I tell them to help me paste together even narrower slivers of Mona, so we can take it to the limit.

It's a painstaking job, and we lose count of how many slivers we assemble. Finally, we have to stop. This is as far as we can go, I tell the Sym team, but you can picture what the culmination of your idea would look like.

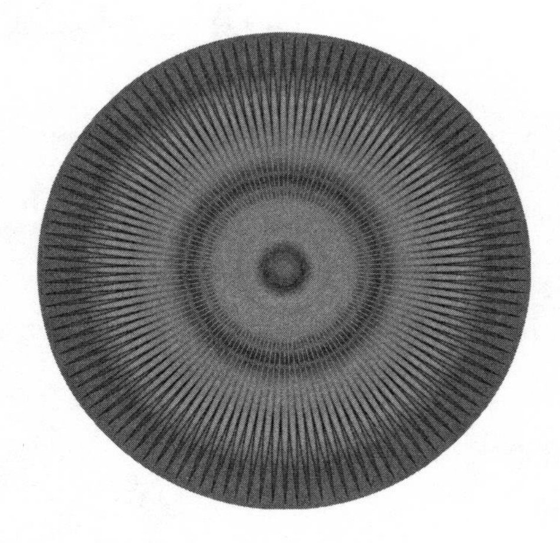

Everyone is transfixed. In Mona's final transformation, she's reincarnated as a circle. This is the limit you arrive at when you keep going the way we did: the most symmetric figure there is. Is it still Mona, though? Yes, in a blended sort of way. Her essence pervades the disk, but it's been homogenized, abstracted.

The Sym team is happy now—it feels vindicated. The Ratio team declares the whole exercise a cheap trick. The two teams can bicker about this forever, but it really boils down to choosing between two different aesthetics. The first is the Golden Rectangle beauty of the original Mona Lisa, the second is the geometric one of this Mona Lisa circle. Perhaps the first is more subjective, something apt to inspire an artist; while the second is something mathematicians might tout as more objective. It really depends on your emotional response, your taste.

This concludes our show, as well as our break. Enjoy your final sips of beer with the pope, as we get ready to join our thought experiment again.

25.

FRACTALS

So far, we've taught Nature basic figures like squares and circles, which she's signaled she intends to follow only approximately. She's also shown she's capable of generating some forms of her own, like the spiral. Is this repertoire big enough to build the universe?

Were we to look at the existing world around us, the answer would be no. Gaze at the sky and you'll see the swirling edges of clouds, view a coastline and you'll observe a series of indentations at every scale, examine a head of cauliflower and you'll notice the billowing shapes florets take. None of these wild boundaries can be easily described by the smooth contours Nature has so far in her tool kit.

So we definitely need more patterns, at least if we take our physical universe to be the guide. Is there a way to arrive at these patterns purely through the numbers instead? After all, that's the aim of our thought experiment. What twist can our narrative take to lead us to such new shapes?

Since the answer has to come from numbers, let's note we've made good use of most of them. Even the irrationals have had a chance to show their worth as the source of randomness. There's one group, though, that's never enjoyed the limelight—and that's the complexes. Sure, they helped us get from one to two dimensions, when we created the plane as real estate to house them. But soon after, they were banished. Why not have them generate the new curves we anticipate?

Integrate them more fully into the universe, give them a more important role to play?

This is particularly significant since the complexes may be unfamiliar to many readers. Where's the payoff for all the effort expended in their construction? We need to show their usefulness, or they'll fail to make a lasting impression. (There's also the matter of guilt. If you recall, we tore the complexes away rather brutally from their rightful condos on the plane. Why didn't we think back then of time-shares?)

Fortunately, everything ties together. The unusual shapes we mentioned emerge naturally once we resurrect the complexes.

The squaring game

To set up the triumphant return of the complexes, let me tell you about—you guessed it—another game! Inspired by the popularity of the Fibonacci sequence, the numbers have been generating new sequences that follow different rules. The most popular of these comes from a game that begins with some initial real, and then applies the rule

$$x \longrightarrow x^2$$

repeatedly.* For instance, starting with $x = 2$ gives $x^2 = 4$. Then this value 4 is taken as the new input x for the formula above, to give the output $x^2 = 16$. Continuing, this gives the sequence 2, 4, 16, 256, 65536, ... In the same way, if we start with $x^2 = 0.5$ instead, the x^2 rule gives the sequence 0.5, 0.25, 0.0625, 0.00390625, 0.00001526, ...

---

* Recall that the Fibonacci sequence reduced approximately to the rule $x \rightarrow 1.618x$ after some terms. So think of $x \rightarrow x^2$ as a variation of this, i.e., as another input/output rule.

Observe that one of two things can happen with this sequence. Start with $x = 2$, and the subsequent terms keep growing. Beyond a thousand, a million, a trillion—the sequence leapfrogs over any bound one might put in its way. However, if you start with $x = 0.5$, then the numbers get smaller in size. They remain within fixed bounds (between 0 and 1, in fact), which is the other type of behavior the sequence can exhibit.

The dividing point turns out to be 1. Starting with 0, 1, or anything in between, and squaring repeatedly, you remain bounded. Anything greater than 1, and you lope off toward infinity.

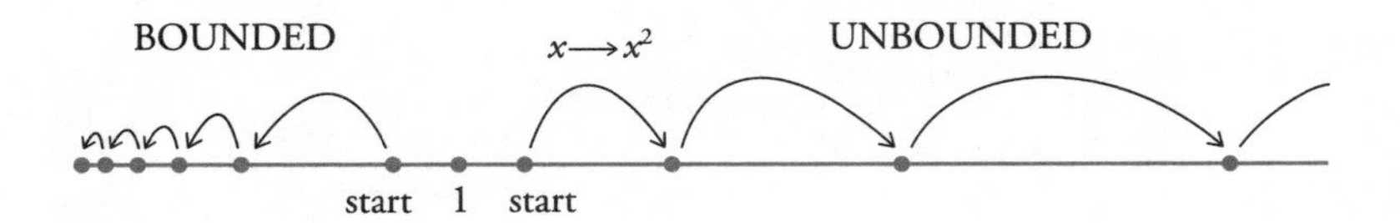

One can start with negative initial values as well. With the first output, x^2, the numbers all become positive, after which they're covered by the previous observations. Anything more negative than −1 again increases unboundedly, whereas starting points between −1 and 1 (both inclusive) remain bounded.

The squaring game is fun at first, but the novelty quickly wears off. There's only so much hopping one can do along the line, which is very constricted. The game gets interesting again only when you bring i back into the fold.

The return of i

Upon seeing i, you feel an instant pang of regret. She seems to have shrunk, her body looks wilted, her face pinched. But exile has not broken her—she's still as tough and strong-willed as ever. Waving aside your apologies, she gets straight to the point. "If you're bored with the squaring game, try starting with one of the complexes. That way, instead of the line, you'll be able to hop across the entire plane."

What i suggests is to replace the rule $x \longrightarrow x^2$ by $z \longrightarrow z^2$, where we've changed the variable x to z, to indicate that z is complex. Recall that complex numbers can multiply each other, just as the reals can, which is why $z^2 = z \times z$ makes sense. There's a formula for such multiplication given in the next footnote, but it's not important for our purposes. All you need to know is that if you take a complex number z as input, then the rule gives a complex number z^2 as output. Moreover, the input can be taken to be any point $z = x + iy$ on the plane, which, recall, is just another representation of (x,y). So as i promised, your playground is no longer just the line, but the entire plane.*

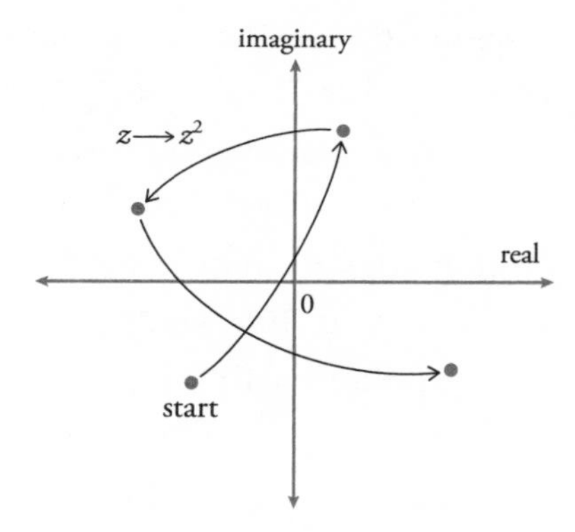

Indeed, start with an initial point z, and you find yourself hopping to a series of points on the plane, each of which is the square of the one before. You can now ask the same question you asked on the line: Will you remain in a bounded region, or will you hop farther and farther away?

---

* Technically, the rule $z \longrightarrow z^2$ can be expressed in terms of reals alone, but the formula is unwieldy, and can be skipped. For those who want to see it, write $z = x + iy$. Then $z^2 = (x + iy) \times (x + iy) = x(x + iy) + iy(x + iy) = x^2 + 2ixy + i^2y^2 = x^2 - y^2 + 2ixy$ since $i^2 = -1$. So $z \longrightarrow z^2$ is the same as saying $(x,y) \longrightarrow (x^2 - y^2, 2xy)$. Notice how the complexes give a much more compact representation.

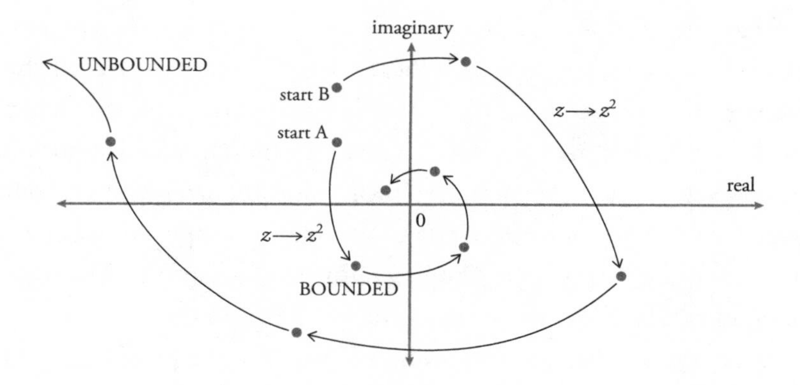

You play the game with different starting points. The first one, A, leads to a bounded progression—bringing you closer and closer to 0, in fact. But then you try another point, B, and this one seems to skip away from 0. You also find that a few points settle down into going back and forth between the same two or three or more locations.

All this is entertaining enough. Plus it soothes the guilt in your soul to bring i back into the fold. But where are the curves you were promised? All you see are points darting around, not the boundaries of cauliflowers or clouds.

A new family of curves

I haven't forgotten about the curves. Here's how to get them. If a starting point leads to a bounded sequence, color it black; if not, leave it white. What picture does this give? You have to check the trajectories of many points to see it, but what will emerge is a perfect unit circle!

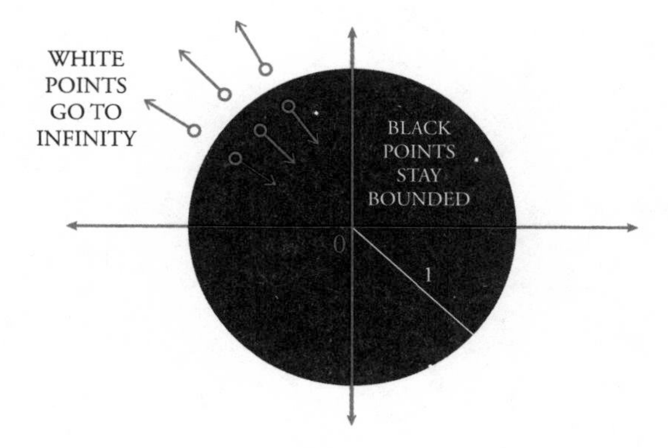

Isn't it amazing? Start at any point in the unit disk, and the iteration remains bounded; start outside this black region, and the iteration becomes unbounded. The border that separates the two regions is a perfect circle!

I can understand if you're less than thrilled. To end up with a circle after all this buildup is an anticlimax. This is hardly the kind of snazzy curve promised.

But we've barely started. To get something less common, tweak the rule a bit. For instance, let's change it with a tiny extra term, to

$$z \rightarrow z^2 - 0.1$$

instead. Notice that this will still give you a trajectory of points on the plane—all you do is subtract 0.1 after squaring z. Once more, start with different points and color them black or white depending on whether they remain bounded or shoot away. You'll notice the border in between is no longer a circle.

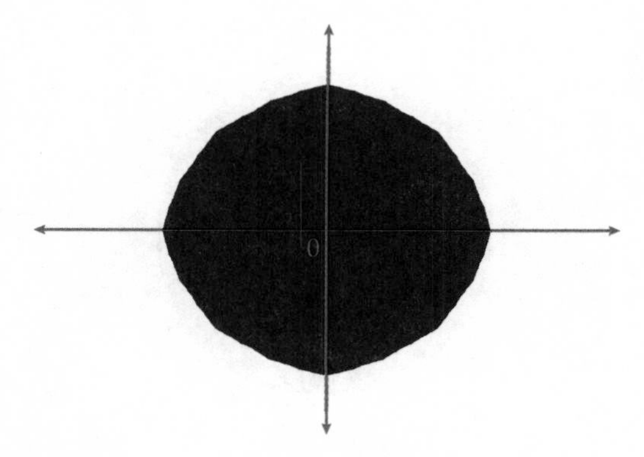

Still not very exciting, is it? Let's tweak things some more. Change the 0.1 to a 0.3, as in

$$z \to z^2 - 0.3$$

and try it again.

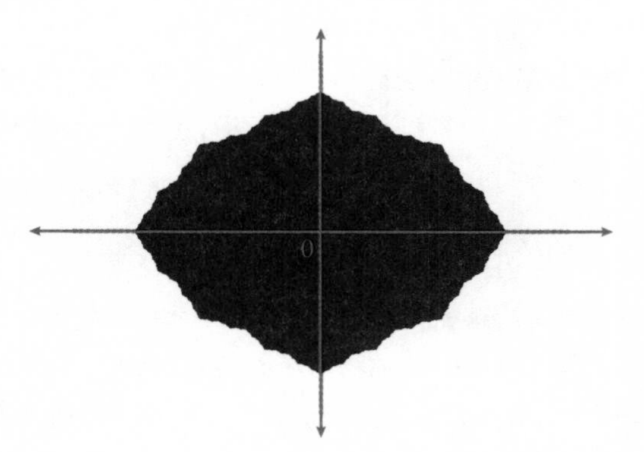

This time, the picture is more interesting. But notice how much more interesting it gets when you change the 0.3 to a 0.6!

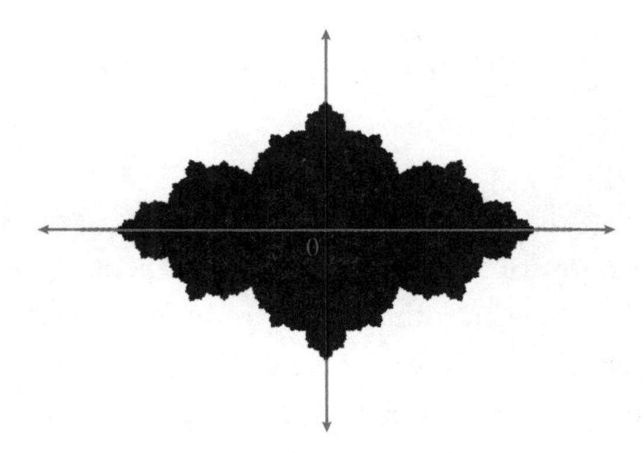

The boundary might now look rather intricate, but it's actually even more remarkable than you might think. Let's magnify a portion of it.

We see that variations of the same shapes repeat over and over again at different scales of the boundary, a repetition that continues no matter how much we zoom in. Each nodule seems to sprout more similarly shaped nodules, which, when we look closer, sprout more of the same.

Imagine walking along this boundary. Would you ever be able to go along a complete nodule? Or would there be so many subnodules that you'd lose yourself in their intricacy, their endlessness? One thing's clear. The boundary is so infinitely complex that it's impossible to duplicate with any curve in Nature's current tool kit.

Of cauliflowers and clouds

The above boundary is an example of a *fractal*. You've probably seen such figures depicted in wild, psychedelic colors—they're eye candy everyone can enjoy. But they're more than mathematical decorations. Even the colors have meaning—they indicate how fast the points outside the bounded black section tend to infinity. As you move away from the boundary, each gradation represents more rapid unboundedness, until the darkest points in the outermost region whiz off to infinity the fastest.

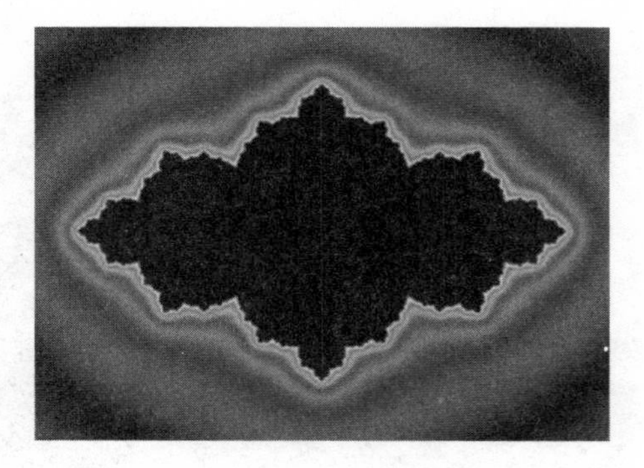

A distinguishing characteristic of fractals is that, zooming in, you see repeating variants of similar patterns. Such self-similarity may be inexact (as above) or exact (as we shall see later). Try measuring the length of the boundary in the above figure and you'll discover that the endless protrusions and indentations make it infinite. Such a border therefore creates a lot more exposure where black can interact with white, compared to regular borders like circles or straight lines.

You can generate a bunch of very different outlines by changing the value of the constant in our generating formula. For instance,

$$z \rightarrow z^2 + 0.25$$

gives a symmetric pattern.

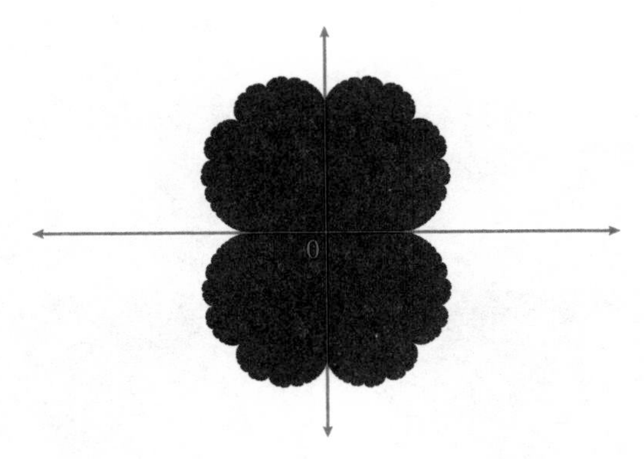

Blow up a section of it, and you start seeing the inspiration behind clouds and cauliflowers. For a cloud, such boundaries represent the interface between moisture in its invisible vapor form and visible condensed form. The fractal nature of such borders expresses how complex the interaction between the gaseous and liquid (or solid) phases is.

A different set of boundaries is obtained when, instead of adding a real number to z^2 in the output, you add something complex. For example, here's what you get when you use the formula

$$z \rightarrow z^2 + i.$$

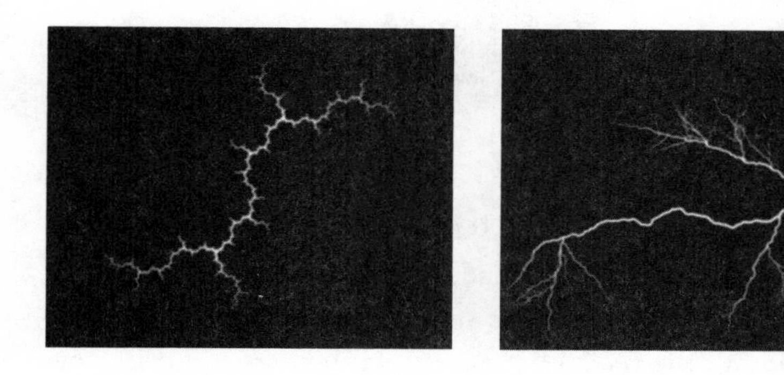

(The white and black have been interchanged in the picture on the left.) Now it's the spikiness that's eye-catching. Could this be stimulation for Nature when she eventually shapes lightning?

Add some other complex numbers instead of i, and you can get fractals that look very much like the eddies found in rivers and oceans. Van Gogh's *Starry Night* might also come to mind.

Looking at the above pairings, there's a stark and inescapable difference. We're familiar with the clouds, cauliflower, lightning, and eddies that exist in our universe, so we know they have none of the order or uniformity seen in the fractals they resemble. This is because of two reasons.

First, the fractals derive from an infinite process, leading to an unending hierarchy of protrusions or spikes. Nature, however, is much too lazy to go beyond the finite—so she constructs an approximation, with a couple of levels of the pattern, and calls it a day. In her defense, she has to use physical particles (of water, vapor, botanical material, and so on) to create these patterns, so it wouldn't make sense to make them too detailed in any case.

Second, Nature's still very much infatuated with randomness. No matter how much our formulas inspire her, she's not going to adhere to them. Her nodules and spikes won't follow any orderly pattern—but will be generated by, so to speak, throwing the dice. Just another example of the imprint she's bent on stamping on our work.

26.

THE WHY AND HOW OF NATURE'S PATTERNS

IF YOU'RE TURNED OFF BY THE EQUATIONS (ALL THOSE z's and z^2's) in the previous chapter, here's a formula-free pictorial way to generate a fractal. For any black triangle you see in the input, remove its middle to get the output.

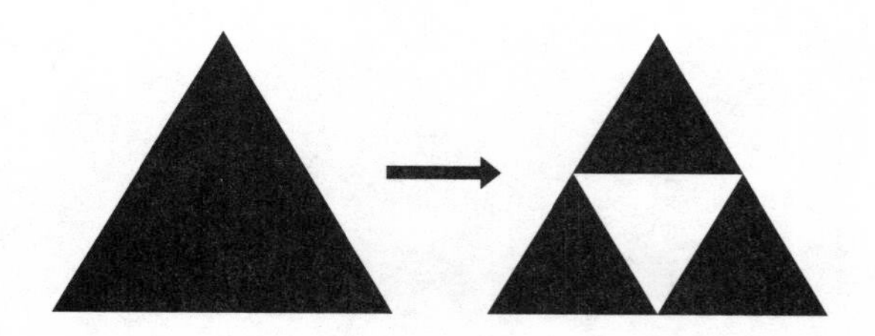

Once you create the output with the three smaller black triangles above, plug it back in as your new input. The rule tells you to remove the centers of each of these three smaller black triangles, which means that there are now nine still smaller black triangles in your new output.

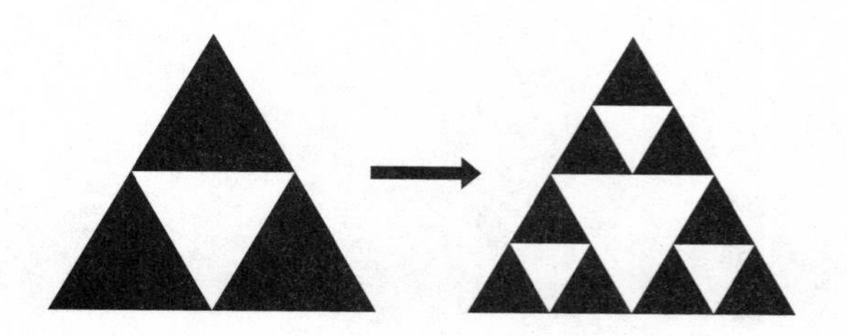

Repeat this a few times, and a pattern will quickly begin to form.

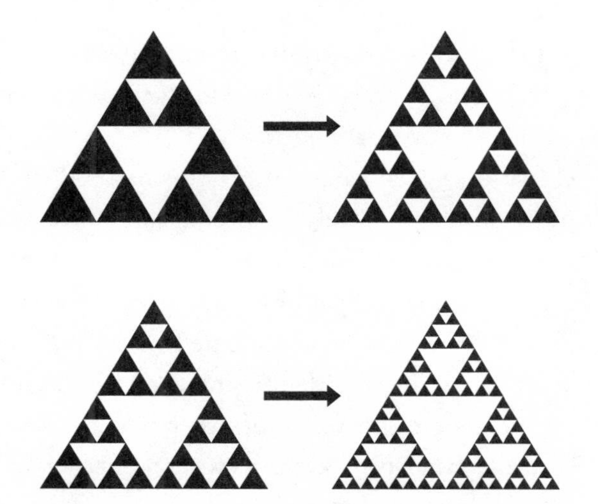

Do this an infinite number of times, and in the limit, you get a fractal (called the Sierpinski triangle) that displays perfect self-similarity. Zooming in gives *exactly* the same pattern as the original, as opposed to a variant in the same vein.

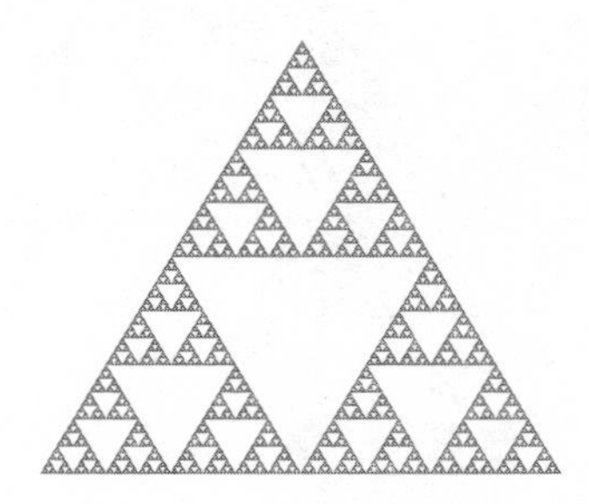

Notice something striking: even the tiniest black region is riddled with holes. Suppose you were able to physically cut all these holes out of a triangle of black paper. Then a particle lobbed at the triangle would pass right through, no matter where it was aimed. There wouldn't be even a tiny scrap of black area left to act as a barrier.

This raises the question: What is the dimension of this figure? Since it doesn't incorporate any area, you can't quite call it two-dimensional. On the other hand, it's clearly not one-dimensional like a line or curve either. Mathematicians address this contradiction by ascribing an intermediate dimension to the figure—a value more than 1 but less than 2. In fact, similar fractional dimensionality is a defining property of *all* fractals (see the endnotes for how this dimension is mathematically calculated).

Of course, Nature, when we teach her this method, is too lazy to create an infinite number of levels. Her rendition of the triangle fractal has only a finite number of holes in it, because she stops after a few iterations. So were she to deploy it physically in our universe to come, she'd have a 2-D pattern, not one with true fractional dimension.

Once you get more adept with this input/output technique, you can use it to generate a rich catalog of other fractals. This includes, for instance, various branching patterns where, once again, the same subpattern is exactly repeated at each level.

Such patterns, once Nature assimilates them, will be the inspiration behind different varieties of ferns and trees. They're also going to underlie other biological structures that involve branching—for instance, lungs and circulatory systems.

Why Nature uses fractals

Let's pop up from our experiment again to ponder another question. From clouds to cauliflower to trees to lungs—there's more than enough evidence that Nature likes fractals. But why? Why doesn't she just stick to shaping everything with smooth curves like lines and circles?

There's no single answer for this. In some cases, the fractal structure presents a clear-cut advantage. For instance, for breathing, it's essential that the network of bronchial tubes have intensive penetration in the compact volume offered by each lung, so that there is maximal surface area available through which air can be exchanged. Fractals are ideal for this; we've already seen how good they are at maximizing boundaries (even managing to make them infinite). The same maximizing effect is relevant for the system of veins and capillaries in our bodies. We need them to form a network that branches out everywhere and facilitates the largest interface with surrounding tissue.

In other cases, fractals may just provide the shortest and most efficient blueprint for construction, given that the same algorithm is followed at every level. We've already noted this as a reason why Nature might be so fond of self-similarity. It's so much easier to use the same set of instructions for fern leaves as you do for their leaflets (pinnas), and subleaflets (pinnules).

A third answer arises from the general kind of recursive rule we used to create the Sierpinski triangle fractal:

$$\text{Input} \rightarrow \text{Output}.$$

Such iterations form the basis of many of Nature's processes, where things that evolve over time undergo a series of transformations from one stage to the next (the output state at the end of any stage becomes the input state for the next stage of evolution).* Since we've seen that the triangle rule leads to a fractal, we can expect that other input/output recursions might as well. A good example is erosion, where you

* Such rules often define so-called dynamical systems, which are mathematical or physical processes that evolve in time.

can think of the input/output stages being freeze-frames at different instants of time. This can result in some instantly recognizable fractals—for instance, the Sierpinski-like pattern carved into a sandstone hillside in Utah shown below.

What's remarkable about this example is that its formation can be explained by the same geometric algorithm that led to the Sierpinski triangle at the start of this chapter. Recall the black areas removed at each step of the algorithm—think of these now as corresponding to material removed from the hill. The difference is that, due to randomness, a mix of big and small sections of material might erode at each stage from the hill, instead of such erosion following the strict progression seen in the algorithm, where bigger triangles are removed before smaller ones. The triangular self-similarity emerges due to a combination of two effects: strata of harder materials resist erosion to form the bases, and the softer sandstone, composed of compacted granules, erodes to create slopes with the same natural angle of repose everywhere.

More subtle fractal patterns are obtained when we consider the effects of erosion on a coastline. The process of removal is similar—water lapping away at the land, random wind gusts sculpting away little bits, perhaps even the occasional tsunami or earthquake causing a major upheaval. The fact that such forces act equally well on microscopic and macroscopic scales is what leads to approximate self-similarity. A surge of water, for instance, can sweep away boulders as surely as it can pebbles or grains—it just depends on the intensity. Wind might whittle away an outcrop, weakening it to a stage where the next tiny gust might make it collapse. Over time, the aggregation of such effects leads to similar features appearing at different scales, as can be seen from the coastline from Florida shown below. Each time you zoom in, you can find a peninsula that looks similar to one seen at the previous scale.*

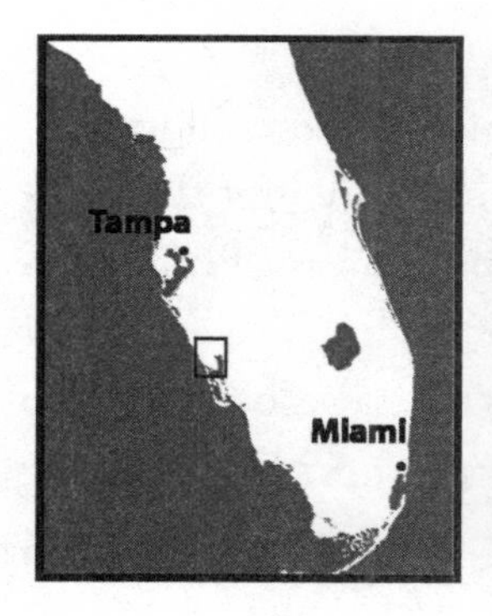

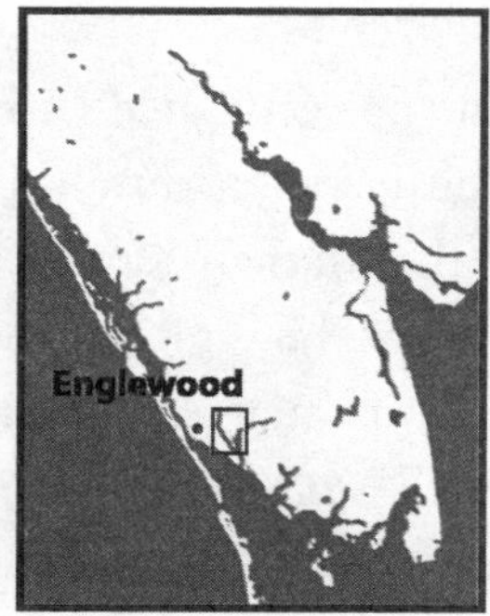

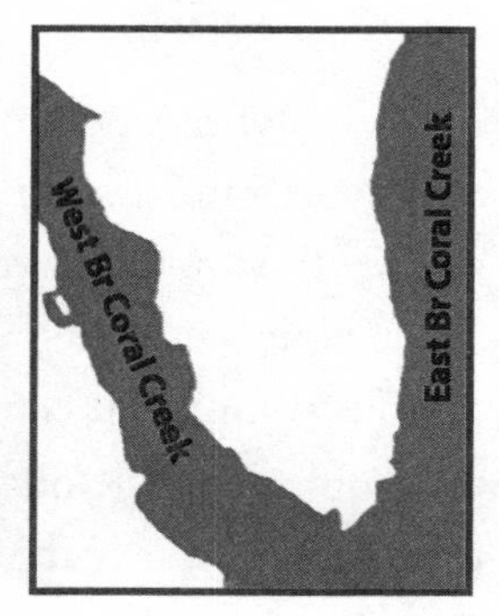

Analogous effects come into play with the topographical elevation that creates mountains and valleys. Nature's agents—water, wind, and so on—will randomly raise and lower spots on the Earth's surface. This gives landscapes a fractal nature—one we might not be conscious of but our eyes get trained to look for. If it's absent, we can tell something is off—mountains might appear flat, valleys artificial.

---

* Due to this fractal effect, it's quite difficult to measure the length of a coastline. The shorter the measuring scale you use, the greater the length will typically turn out to be, because you can now measure all the indentations at the finer level that you couldn't before.

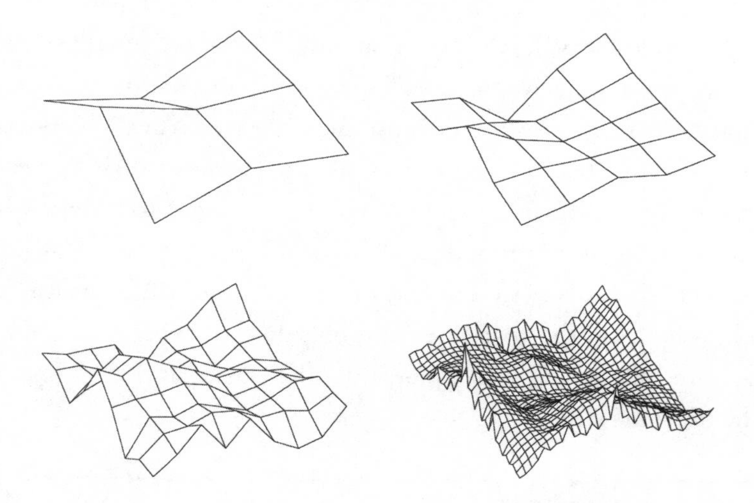

Such landscapes can be generated using random numbers. Although a more sophisticated algorithm has been used for the landscapes illustrated above, the simplest idea would be this. Pick a random sequence of digits—say the digits in the decimal expansion of pi.* Partition your area into squares, and either raise or lower the midpoint of each square depending on what the next digit is. (For instance, raise it if the next digit is 0, 1, 2, or 3, lower it for 4, 5, 6, 7, and keep it the same for 8, 9.) Then divide each square into four smaller squares and continue the process. Numbers, it turns out, can be just as effective agents of erosion as water or wind.

The ultimate vindication of this effect comes from Hollywood. Ever since the second *Star Trek* movie (way back in 1982), they've been using fractals to make computer-generated terrains more realistic. Chances are you'll find fractals embedded in every sci-fi landscape now: movie, video, or computer game.

---

* This would be only pseudorandom, not random, as previously discussed. Also, you'd probably want to use the expansion not from the first digit but from some other randomly picked starting point, say the 5,386th digit.

Chaos

There is a deeper, more hidden reason why fractals occur in nature. They are often an indicator that some underlying physical process being carried out by Nature is susceptible to *chaos*. One of the defining characteristics of a chaotic process is that small changes in the initial state can create large changes in the resulting outcome. The classic example is the "butterfly effect," said to occur when the flapping of a butterfly's wings is enough to set off a tornado several days later on the other side of the world. Think of the weather as an input/output rule, where the current atmospheric conditions (fed in as input) give rise to the atmospheric conditions a few instants later (the output).* A tiny change in input (the butterfly flapping) can get progressively magnified through successive input/output cycles to produce an enormous change in the output (the tornado). In fact, the butterfly effect is why it's so hard to predict the weather more than a few days out. It's impossible to nail down every bit of weather data exactly, and even the slightest inaccuracy fed into a meteorological model can (and often does!) cause a major error when the forecast for a week or so later is calculated.†

The mathematical input/output rules (such as $z \to z^2 - 0.6$) used to generate various fractals in the previous chapter can exhibit the same kind of sensitivity to initial data as the input/output rules for weather.

---

* Other data, such as land temperatures, topography, sea conditions, etc., would also have to be part of the input and output.

† This fact was accidentally discovered by the MIT meteorologist Edward N. Lorenz in 1961. He found that a tiny change in input data (using the number 0.506 instead of 0.506127), when fed into a simple computer weather model he'd designed, resulted in a completely different long-term forecast. Since such approximations are unavoidable (data involving temperature, rainfall, and so on, cannot be measured very accurately, and so has to be input in approximate form), long-term weather prediction is essentially impossible. The term "butterfly effect" originated in a talk he gave in 1972 to the American Association for the Advancement of Science, entitled "Does the Flap of a Butterfly's Wings in Brazil Set Off a Tornado in Texas?"

Recall that we colored a point black if it remained bounded under repeated application of the rule, and white if it went off toward infinity. Think of what happens near the boundary between black and white. A tiny change in where you start—the black area versus the adjacent white area—leads to a dramatic difference in your fate. You either end up somewhere bounded (the equivalent of good weather) or at infinity (perhaps transported there by a tornado, like Dorothy in *The Wizard of Oz*). Since the boundary between the two different outcomes (i.e., between black and white) is often so complicated that it's a fractal, it becomes very difficult to control your fate.

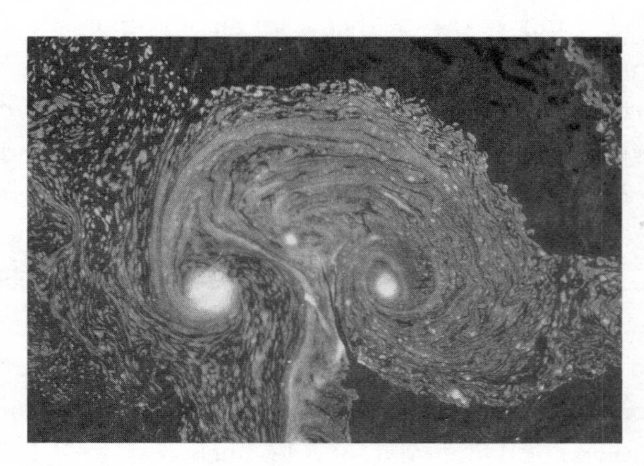

Such fractal boundaries are common when an input/output rule sensitive to initial conditions is at play in nature. You might already be colloquially using the term "chaos" to characterize some such situations—for example, when turbulent water flow creates eddies and vortices. In the image above (similar to one seen earlier), the black areas are regions of relatively stable flow, where water moves in orderly parallel layers, with little intermixing between layers. In contrast, the white areas represent regions of turbulence, where particles get pulled into vortices, and layers vigorously swirl and mix together. Notice how large and complicated the interface area is, where black and white commingle. This area acts as a kind of catch basin: start there, and you'll probably be sucked into turbulence.

Interestingly, cloud boundaries, which separate air (where water is in vapor form) from the cloud (where water is in condensed or crystallized form) exhibit fractal character for a similar reason. The dynamics of water vapor both inside and outside the cloud can get quite turbulent, once again leading to a chaotic process.

What about the weather? We know its underlying input/output processes are chaotic, but does this get manifested in fractal behavior? Indeed it does—the variability of meteorological phenomena (e.g., rainfall, temperature, wind velocity), in terms of both geography and time, is often fractal in nature. Look at a map of rainfall distribution that carefully accounts for local microclimates, and you will see that the boundaries of different zones have the same fractal character as the coastlines we discussed earlier. View a graph of rainfall plotted against time, and you are likely to notice that jagged day-to-day fluctuations are similar to the larger fluctuations observed year to year. As in other processes, like erosion, Nature takes care to smudge any self-similarity with a good dose of randomness.

The shell game

Whatever her specific motive, when Nature creates a fractal pattern, it's often an indication that she's using mathematics in the form of an input/output rule somewhere. Such usage might not be apparent, in which case it can be a challenge to unearth the rule and model it. As an example, let's consider the way the Sierpinski triangle shows up prominently on the surface of certain seashells.

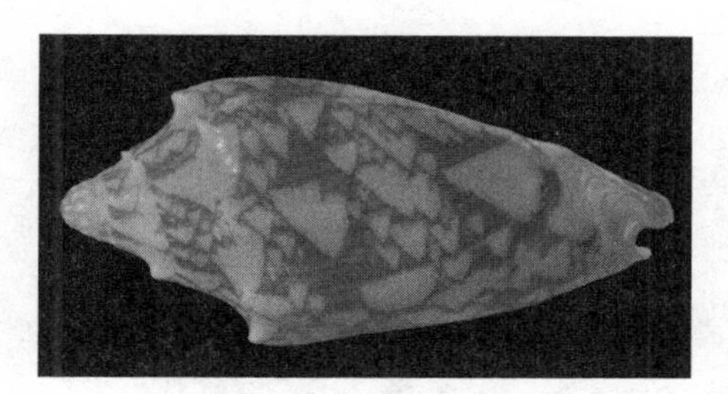

Take a look at the above pattern. How might it have come about?

It couldn't have appeared through erosion as it did on the Utah hillside—neither wind nor water could do the trick here. No analog of the geometric algorithm could work here, since a mollusk could hardly white out sections of its shell once generated. So what exactly did Nature do? Delving into this question should give us more insight into how she wields mathematics.

Mollusks create their patterns through a row of pigmentation cells on their *mantle*, a tongue-like protrusion that curls around the outer ridge of the shell. These cells turn "on" or "off" while the mantle secretes new material along the shell's growing edge.

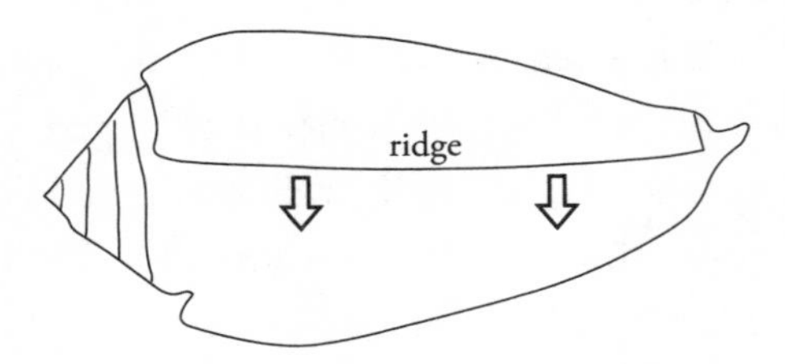

Some mollusks keep the same fixed selection of pigmentation cells activated indefinitely (with the rest off)—this leads to stripes perpendicular to the outer edge. Others switch the entire row of pigmentation cells on for isolated bursts of time and leave them off otherwise, creating stripes parallel to the edge.

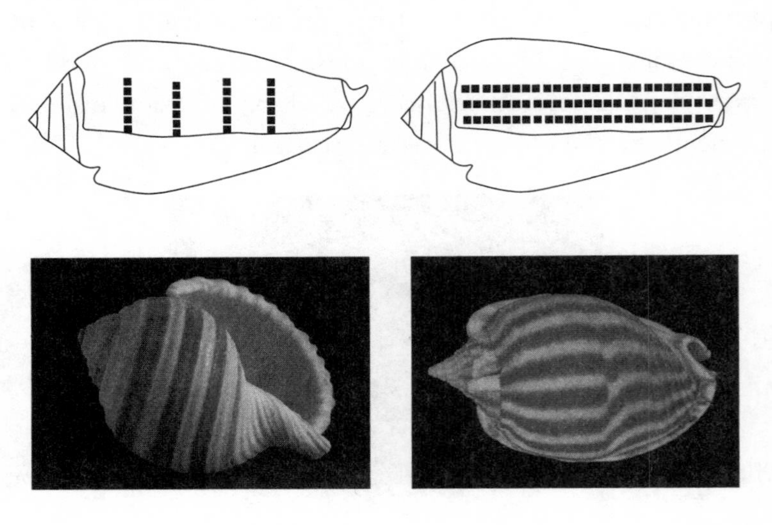

But what if a mollusk were to switch its pigmentation cell on or off depending on what this cell, along with the adjacent cells on either side, did in the previous step? This is the proposed mechanism we will follow. Think of this as an input/output rule with the input coming from three inputting cells.

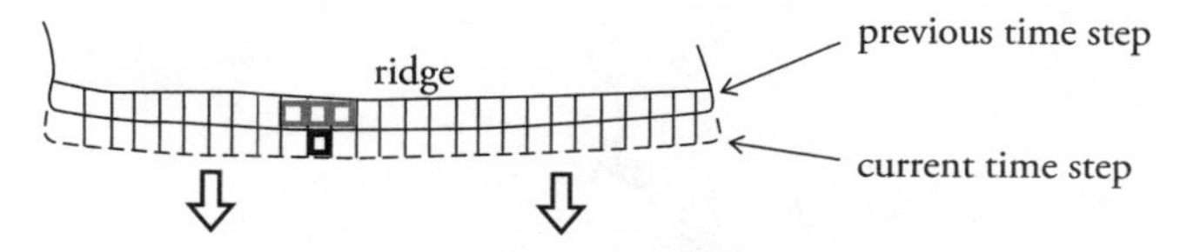

Pigmentation of 3 top squares determines that of bottom square

Consider the following rule. The cell switches on at the current step if either one or two of the three inputs received from the previous step were on. However, if *all* three inputting cells were on, then the cell switches off—perhaps because too much pigment is being used. Finally, it also remains off when all three inputting cells were off. This array of possibilities can be represented as follows.

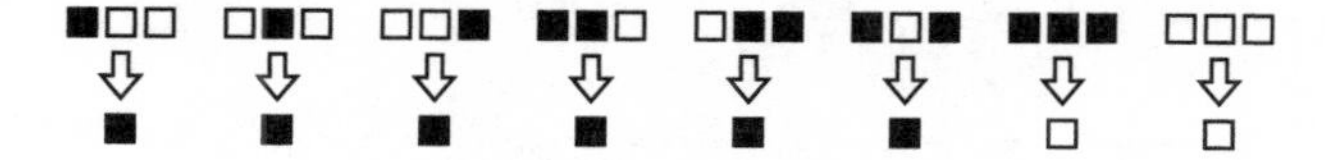

Now suppose you start with a single pigmented square, as in the shell below. To see how the coloring would progress, fill in the squares for each new row according to the rule above. What do you notice?

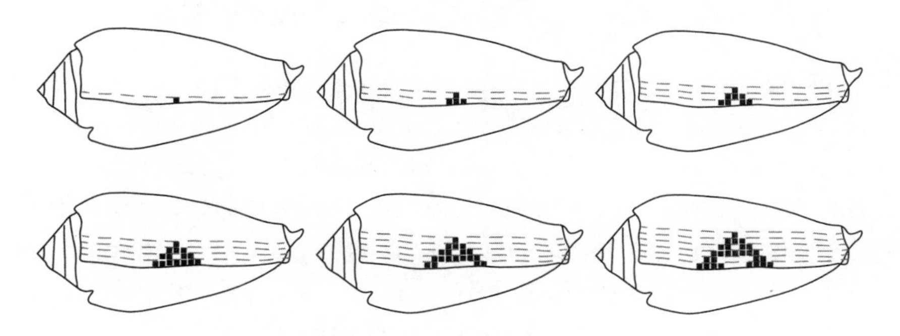

Yes, it's the Sierpinski triangle fractal beginning to emerge! Continue for some more steps, and you will see it in all its glory. The simple three-celled rule has led to a pattern with great complexity!

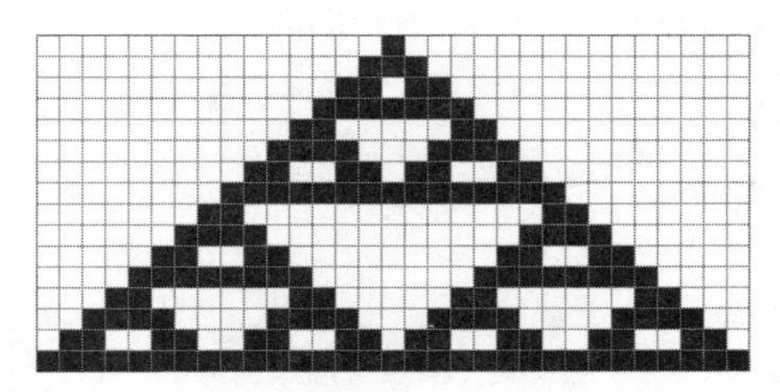

The above is what we get when we start with just one black square. But in an actual shell, the initial point will be some random distribution of pigmented cells. Then the pattern you get will also have a lot of randomness in it. An example is shown below, where the topmost row now has several pigmented cells, not just one. Observe that you do, indeed, come up with the kind of pattern seen in the actual shell displayed earlier.

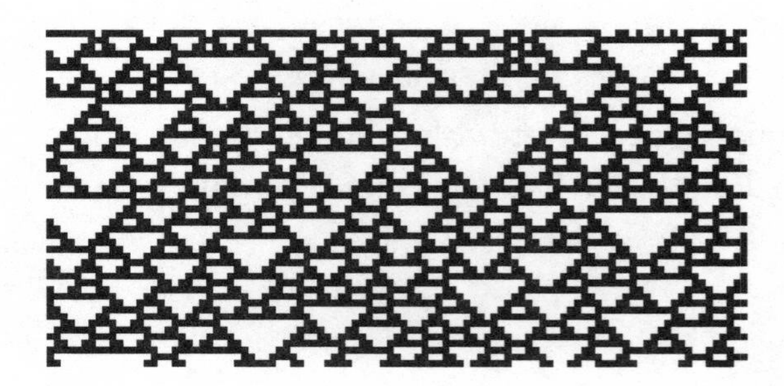

Could there be other input/output rules? Look back at the rule's characterization, and you will see that each of the 8 outputs can be either black or white. Changing even one of them gives a different rule, with possibly a different pattern. There are $2 \times 2 \times \ldots \times 2 = 256$ different rules you can formulate, so it's no wonder you see so many different designs on shells. In this sense, the Sierpinski pattern seems to be one of many that's just waiting to emerge.

So how does this fit into our picture of mathematics and Nature? We could say that the rules above are something we've invented to explain the behavior of the mollusk. This would be true enough. But notice that we could also think of Nature figuring out a way to create the fractal pattern we gave her (in her own approximate and random way) by delegating the task to pigmentation cells. In fact, think of her sitting on a beach somewhere, trying out all these different rules, delighting in all the differently patterned shells she generates. Spots, stripes, fractals—perhaps she purposely adds a few errors in each pattern to mark her imprint. The exhilarating thought is that she's taken to heart the idea that mathematics is all about play. What could be more satisfying than the universe emerging the way it did due to games?

One last surprising pattern

You might think that Nature's penchant for randomness always works against the patterns you've so painstakingly invented. That it invariably throws orderly geometric designs into disarray, making them less recognizable. Imagine your surprise, then, when you discover that the opposite can be true as well. Using randomness, Nature can actually *build* patterns you intended.

She shows you this herself one day. She's assembled a band of particles—the more obedient ones, she says—whom she's managed to herd together in a tight little knot. They're buzzing impatiently, turning this way and that, but following her order to stay together. "The key is how they all face random directions," she tells you. "Watch what happens."

She addresses the particles. "At my command, step forward exactly one unit." She pauses for a moment, and you can feel the tension grow in the knot. "*Now*," she says.

A flurry of activity ensues as all the particles move forward. What emerges is an almost perfect sphere of radius 1. Due to the mix of directions in which they were faced, the particles end up distributed all along the sphere's surface.

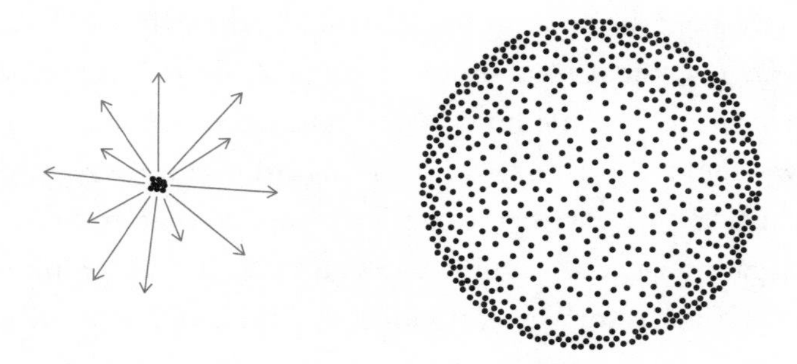

You stare, dumbstruck. Instead of using algebra or geometry to generate a sphere, Nature's just used randomness! A feat accomplished without the help of the coordinate system, without lighting up any points. Your mind reels at the implications. Will she jettison all the mathematics you've set up?

You're about to confront her with this when you have another thought. She's still created a sphere, a mainstay figure your geometry has given her. It's only her technique that's left you perturbed. But surely the method of construction is her prerogative, as it would be for any contractor. If anything, you should be happy she's so resourceful. Moreover, the idea of randomness in your universe stems from the irrationals (via the unpredictable digits of pi or $\sqrt{2}$, for example). So she's still using mathematics that derives from your numbers.

You realize what's really bothering you is how Nature's reducing her dependence on you. All the tricks she's been coming up with, the muscles she'll be able to flex soon. But the universe still has a long way to go—once things become physical, it's going to be much more complicated than anything she can handle alone. She'll need your help, or at least that of mathematics, to lay down the laws of physics. No matter how independent she gets, math will still exert control.

Day 5

PHYSICS

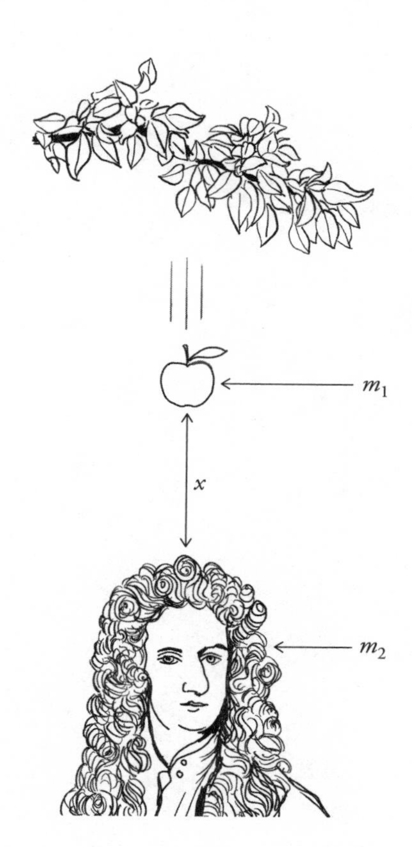

Helping Nature develop physical laws

27.

A UNIVERSE RUN BY LAWS

It's time to begin planning the working of our physical universe. Specifically, how things, when jelled, will interact with one another. Let's start with Nature's simulated particles. If these are left on autopilot, what laws would ensure their orderly behavior?

There's a philosophical question we must first negotiate. Recall that Nature, depending on your view, can be powered by God or physics. While we're all familiar with science having a system of laws at its core, there's no reason God needs to operate the same way. A divine entity could do anything she wanted—make things appear or disappear, stop heavenly bodies in their tracks, create tunnels through time and space. She could rain down calamities when angered, shower gifts when won over. Why would she ever agree to follow any laws? Why not personally perform every task, with the ability to throw in miracles at whim?

The pope has spent considerable time thinking about this. He remembers, as a young student, being awestruck when the chemistry teacher mixed two colorless solutions together to create a brilliant yellow precipitate. Clouds of iodide appeared out of nowhere, surging and billowing through the liquid. It was spectacular, magical—like a miracle from the Old Testament, reduced to test-tube scale.

But then his teacher wrote down a chemical equation: potassium iodide plus lead nitrate gives potassium nitrate and the insoluble lead iodide precipitate. How simply and completely this expressed what they'd witnessed. Surely God would also be won over by such succinctness. If He sanctioned such laws, there'd be no need to perform miracles, or even get involved.

The pope remembers following the thread of this argument. With so many more pressing things to do, wouldn't God want to let chemistry run on its own? Not to mention other forms of science, and engineering, and medicine, and so on? Equations, so indispensable for humans, could be just as useful for the Creator. The fewer times God intervened, the more each intervention would count. Miracles had to be exceedingly rare, or they would lose their power to transform.

So the pope is on the same page as the position we'll take here. Nature's going to need laws, whether you believe she's a manifestation of God or physics or both. She'll let things run by themselves, rather than constantly inserting herself into the mundane.

Mollifying the physicists

That little intro may have rubbed some physicists the wrong way. They feel I'm giving them short shrift, that I've been paying too much attention to religion and not enough to science. Witness the platform I've afforded the pope—where are *physicists'* balancing viewpoints? They're unwilling to keep silent after my transgression today. (They object to such metaphorical "Days" as it is—couldn't I have numbered the sections without bringing the Bible into it?) Look at the heading of this section, they say. It's *physics*—why then does it start with the pope's pontifications? Why not an introduction from Einstein or Newton?

In response, let me first point out there's very little friction between science and mathematics. I may joke about disagreements, but physics and math are really two sides of the same coin. It's religion that has historically been in conflict. That's why I keep soliciting the pope's take on things—it's more revealing. As far as the physics viewpoint goes, I've already been promoting it. All the context I provide, all the explanations I give—I *am* the scientific mouthpiece.

The physicists are still grumbling, so let me promise this. Before this Day is up, we'll give a nod to both Newton and Einstein in our narrative.

The double life of formulas

Let's get back to the laws. It's great we're in agreement on the need for them, but what should these look like? How do we express them? Turns out we can make use of the language we've developed—the language of algebraic formulas.

So far, we've used these just to designate curves. But they're versatile enough to describe rules and relationships as well. For instance, $y = x^2$ represents a parabola, but it can also be pressed into service to express the area y of a square with side x. Similarly, $y = x^3$ is a cubic when drawn, but it also gives the volume of a cube with side x. In other words, equations we know well can be repurposed to define laws.

Now, we've developed a number of formulas—called polynomials—by letting symbols play addition, subtraction, and multiplication games with numbers and each other. Are these varied enough for all the laws to come?

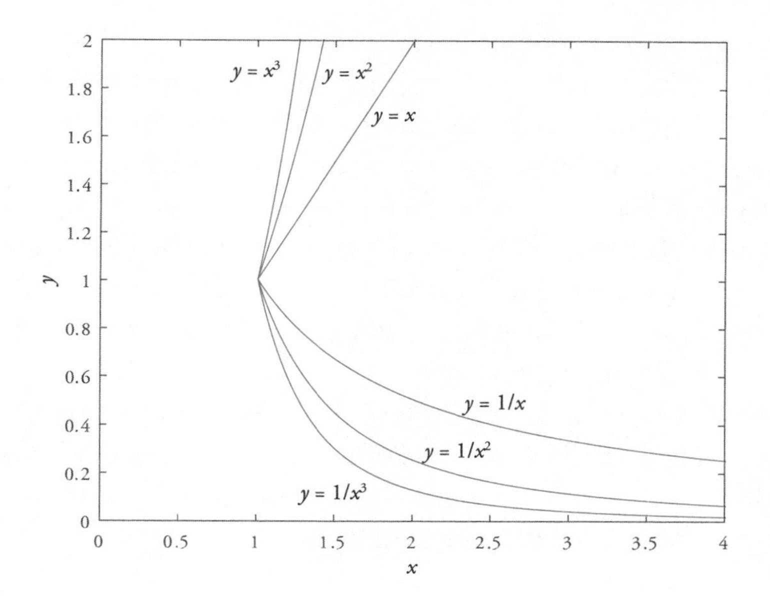

They're not. You'll notice we didn't include division in the above list of games, which we'll now remedy: one set of new formulas we'll need soon is obtained by dividing 1 by powers of x. This gives variations like $1/x$, $1/x^2$, $1/x^3$, . . . that *decrease* with x (rather than increase, as do x, x^2, x^3, . . .).

Let's have the symbols invent even more games to create further new formulas—for instance, by taking roots of x. In fact, entire polynomials can start playing such games—dividing each other, taking roots of themselves, and so on. Once such play has equipped us with a large-enough catalog of formulas, we can begin setting up the universe's laws.

The arbitrariness of our universe

Since we're starting with a clean slate, and have a bunch of formulas at our disposal, it would seem we have complete freedom in our thought experiment regarding which formula to deploy where. But do we really? Perhaps not, if we want to take into account some commonsense principles. Not to mention Nature's reluctance to work more than she has to.

For instance, one of the first things on our list is to ascribe a new attribute to particles called "mass" that will be one measure of how physical they are, once Nature jells them. We don't have a complete sense of this characteristic yet, and so won't try to describe it further, but it's something we'll need to track if these particles are to be building blocks. The idea will be to combine particles to build structures with more and more of this mass. What rule should such combining follow? If particle 1 has mass m_1 and particle 2 has mass m_2, what mass m should their union have?

The choice $m = m_1 + m_2$ seems obvious. We're not sure yet what it would mean for a pair of particles to combine, but it sounds very much like addition. Hence it makes perfect sense for the numbers m, m_1, m_2 to follow the usual rule of addition.

But what if "the whole is more than the sum of its parts," as the phrase goes? What if particles, combined together, result in a mass different from the sum, say $m = 2(m_1 + m_2)$ or $m = (m_1 + m_2)^2$ or even $m = m_1 \times m_2$? Could these alternative formulas work? Could we design a universe where masses combined according to one of these different laws?

This might seem exotic, even outlandish, but maybe that's only because we're unaccustomed to such rules. Keep in mind that all these formulas for m arise from operations between numbers (such as doubling, squaring, multiplying) that are as routine as addition. So perhaps we should explore these possibilities, each of which would give rise to a different universe.

Here's one elementary test we can use to check out these rules. Say we have a particle of mass m_1, and we don't combine it with anything, just leave it as is. We can think of this as combining it with a phantom particle of mass $m_2 = 0$. Now, such a union, of mass m_1 and nothing, should surely just result in mass m_1 (i.e., the final answer m, which is the "sum" of m_1 and 0 should just equal m_1). But look at what the above rules lead to. The first one gives $m = 2(m_1 + 0) = 2m_1$, which means it doubles the mass. The second one gives m = $(m_1 + 0)^2 = m_1^2$, which means it squares the mass. The third one gives $m = m_1 \times m_2 = m_1 \times 0 = 0$, which means it makes the particle vanish! In other words, none of these rules make any sense.*

There are other rules one can cook up, but they're going to have their own defects. If we seek something simple yet sensible, then the usual additive rule is it.

---

* The rules suffer from other defects as well. For instance, one can check that if either of the first two rules is used to combine three particles, then we can get different answers depending on the order in which the particles were combined, i.e., $(m_1 + m_2) + m_3 \neq m_1 + (m_2 + m_3)$. This contravenes one of the basic properties of addition (called "associativity"), which says $(n_1 + n_2) + n_3 = n_1 + (n_2 + n_3)$ for any numbers n_1, n_2, n_3.

Setting the path for Nature

The above was a rather elementary rule, so let's consider something a little more complicated. Once again, we want to gauge how much freedom we truly have in terms of selecting an arbitrary formula from the vast library we have.

We mentioned the idea of using Nature's particles as building blocks. For this, we're going to need some sort of mechanism to pull particles together—something that can create larger bodies through agglomeration. What formula should we select to govern such attraction? In particular, for two given particles, how should the attraction vary with x, their distance apart?

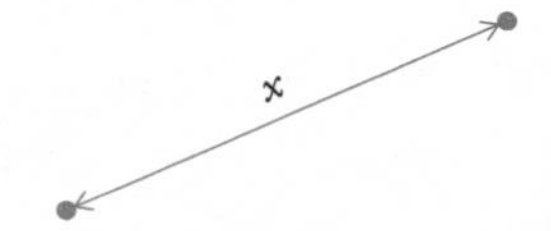

Since simplest is best, why not take the formula to be linear, say x itself? Or some multiple of it, like $2x$ or $3x$?

This idea turns out to be spectacularly unsound. For particles far apart, x would be large, which means they'd experience a strong force of attraction. Consequently, things would start flying toward each other from the remotest corners of the universe. However, as two particles neared, x would decrease, causing their mutual attraction to slump. In fact, if they touched, x would be 0, which would mean they'd have zero cohesive power. Instead of hanging together, chances are they'd zoom off to some different far-flung corners, attracted by new particles there. And with a formula like x^2 or x^3, such effects would be even worse.

Now, maybe such a freakish universe could be designed, but as far as what we want in our thought experiment, objects very far apart should *barely* attract each other. In other words, the attraction should *decrease*, not increase, as the intervening distance x becomes larger. This is the kind of behavior induced by $1/x$ rather than x. So perhaps we should choose a formula where x, x^2, or some other power appears in the denominator.

But *which* power? Do we have the luxury of choosing any of them from the grab bag of possibilities available? Or are we more constrained?

Before we answer this question, let's note how pleased Nature is with the way we've framed this attraction. The fact that we're assuming it's dependent on x means we're only considering laws that are symmetric in every direction (as opposed to a law, say, for which the attraction is more toward a particle to the left than toward a particle to the right). In other words, the attraction a particle exerts will be the same at any point on a sphere of radius x drawn around the particle.

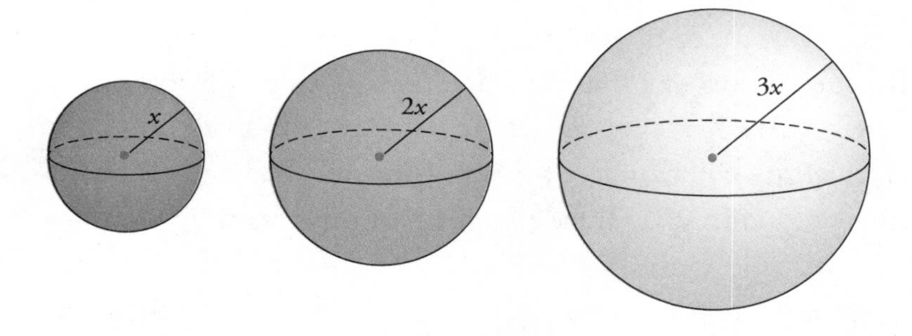

Nature would like us to go further, though, in keeping with her desire for uniformity throughout the universe. Could we therefore ensure that the *total* pull exerted on any sphere around the particle be the same, no matter its radius (x, $2x$, $3x$, whatever)? Her request might sound peculiar, but a couple of analogies will illustrate that it actually makes sense.

First, imagine replacing the particle with the nozzle of a can of spray paint. Suppose this nozzle sprays paint evenly in every direction. Then the inner surface of a sphere of radius x would work just as well at capturing all the released paint as would that of a sphere of radius $2x$ or $3x$ (for which the paint would just travel farther before encountering an obstacle). Replace the nozzle with a light bulb that emits light evenly in all directions, and the same will hold true: a sphere of *any* radius will intercept all the light. Nature's request is that the same be true for the particle's attraction—in other words, a surrounding sphere should "capture" the same total inward pull, no matter what the radius. This certainly is a very reasonable request, one that follows from simple geometry. (Besides, if it doesn't hold, Nature might drown in the bookkeeping involved.)

What will differ, depending on the size of the sphere, is the *intensity*. Since the same amount of paint is being spread out over a larger area, a sphere of radius $2x$ will acquire a much thinner inside coat of paint than one of radius x (for radius $3x$, the coat will be even thinner). The same will be true for the light intensity, and for the intensity of attraction due to the particle.

Now, thinking of the spray can, intensity is simply the total amount of paint deposited divided by the area of the surface that receives it. The surface area of a sphere of radius x is given by the formula $4\pi x^2$, which means it is proportional to x^2. Since we divide by this area, the intensity will therefore be proportional to $1/x^2$. In other words, a sphere of radius $2x$ will have a fourth of the intensity of a sphere of radius x, while one of radius $3x$ will have a ninth of this intensity.*

The attraction a second particle will feel depends precisely on this kind of intensity, calculated in a similar way. If you double the distance between two particles, then the attraction between them will be one-fourth the original; triple it, and it will be one-ninth. We have our answer! The only power of x that works if Nature's request is to be honored is $1/x^2$.

Therefore, the formula we choose cannot be something arbitrary. Our hands are tied by mathematics.

---

* This squaring factor appears because, for example, $\frac{1}{(3x)^2} = \frac{1}{9x^2} = \left(\frac{1}{9}\right)\frac{1}{x^2}$.

The full law

Let's follow this law further, and ask what else this attraction should depend on. Again, math provides the answer: the attraction should be proportional to m_1, the mass of the attracting particle.

Here's the reasoning. Suppose instead of only one particle doing the attracting, we had *two* of them at the same position, and they were identical in all respects. The most reasonable expectation is that we should then get twice the attracting force. But we can think of this situation as the *combination* of the two particles exerting this doubled attraction. In other words, doubling the mass doubles the attraction, and an analogous relationship would hold if mass was tripled. Which means the attraction must be proportional to m_1, the mass of the attracting particle. Putting this together with the dependence on x, we see the proportionality must be to m_1/x^2.

But suppose now that a particle with mass m_2 is the one feeling this pull. Note that m_1 and m_2 can be interchanged, since the particles mutually attract each other—designating a "puller" and a "pullee" is arbitrary. In other words, the law must be symmetric as far as these two quantities are concerned. From this, we're led to conclude that the proportionality should be to $(m_1 \times m_2)/x^2$. In other words, two particles attract each other with a force that is directly proportional to the product of their masses and inversely proportional to the square of their distance apart.

This is, of course, the celebrated gravitational law of Newton—whom we've now encountered, as promised. He supposedly discovered it by watching an apple fall from a tree, maybe even conk him on the head (not true, he says). We've been enlightened as well, and it's by fruit from the same tree—the tree of mathematics.

28.

TIME AND SPACE

THE LAW FOR GRAVITATION IS JUST ONE OF A MULTItude of rules we need to formulate for a working universe. The $1/x^2$ dependence, imposed purely by geometry, will pop up in some of these laws—for instance, the ones governing the intensity of light or spray paint (once we invent light and spray paint, that is). It will also hold for the intensity of various other phenomena still to come: sound, radiation, electrostatic attraction, and more (yes, we have a rather long to-do list). We'll require all these phenomena to act equally in all directions just like light and gravity, with the result that their dependence on distance x will be $1/x^2$.

Other phenomena will exhibit other dependencies, again forced by basic mathematics. We're going to have to design an enormous package of contents for our universe, along with rules and attributes to govern them. At least some laws will have multiple viable formulas from which we can choose. However, much of this latitude will vanish once we start adjusting things to make all our interlocking components consistent. It's one thing to formulate individual rules for everything in the universe, quite another to make them all work in harmony.

Is such harmony achievable? Remember, we can't just work as physicists do, and use observations and experiments on a preexisting universe to figure out our laws. Our thought experiment requires us to find a *mathematical* path to such laws. What would such a harmony-creating mathematical strategy even entail?

The answer is the axiomatic method! This is what lies at the core of math, after all—setting up assumptions that are reasonable enough to be "self-evident" (e.g., "given any two distinct points, there exists a line between them"), then using these axioms to build up an entirely consistent set of mathematical laws. What if we do the same for physics? Figure out the most basic, self-evident principles that govern the universe and use these as axioms to build all its physical laws? As long as our starting axioms don't contradict one another, neither will the deduced laws. So this would give us the harmony we seek, in addition to checking off the "mathematical path" requirement.

Look at the previous chapter, and you'll notice we had already begun to follow this approach. For example, we used the "self-evident" assumption $m + 0 = m$ (an axiom, in other words) to argue that unconventional laws for summing mass wouldn't work. Also, we decreed that gravity had to satisfy conditions of symmetry and "total pull" (again, intuitively evident axioms) so that we could derive the inverse square dependence. The full law involving masses also depended on axioms—for example, that two identical particles should have twice the attraction of one.

In 1900, the influential German mathematician David Hilbert declared the task of reframing physics in terms of such axioms to be one of mathematics' greatest open challenges. The practical role axioms played would now be somewhat different, since science, being based on experiments, described what was already there, and didn't typically lay its foundations first (see the endnotes). Rather, Hilbert's idea was to "reverse engineer" the axioms, i.e., search for a set of them from which the physical laws that experiments had established could be deduced. Although physicists by and large rejected this plan to axiomatize their subject (expressing skepticism that still runs strong today), some theoreticians began working to fulfill Hilbert's ambitious charge. The program remains ongoing, with significant ground having been covered.

Let me point to this program as supporting evidence for the idea that physics *can* successfully be built up as part of our mathematical thought experiment. In fact, let me go further by outsourcing most of Day 5 to Hilbert's heirs, who, in their own good time, will try their best to ensure our to-do list of needed phenomena and formulas is taken care of. (A finesse, I'll admit, but how else to circumvent this massive effort!) This frees me up to present a few more examples of how mathematics shapes physics, in order to bring out more of math's hidden orchestrating power. One of these examples will involve symmetry again; the other will showcase non-Euclidean or "curved" geometry, which we introduced on Day 2. Such geometry, as we'll see, can play an essential role in fashioning the universe's fabric.

A matter of time

To proceed, we'll need to home in on time a bit. So far, we've taken it for granted—something ticking away silently in the background. How exactly do we want it to behave?

The most mathematically straightforward way is for it to progress evenly, neither slowing down nor speeding up, with no skips or stops. This matches well with how we experience it in our universe. We can give it its own axis, with initial point 0, and its own new variable t. This axis will have equally marked intervals along it, counting t in some appropriate time unit (seconds, minutes, hours, nonmetaphorical days, whatever). We'll call any point along this t axis an *instant*.

Where should time start? Our personal experience doesn't stretch back far enough to guide us. One possibility would be at the beginning of one of our Days (perhaps when numbers or geometry or Nature first emerged).* But we could also assume this axis extends

* A more definite choice, based on cosmological calculations, is 13.8 billion years ago, since that's when the big bang occurred.

indefinitely in the negative direction, which would mean that time has existed forever. Then the location of the origin becomes somewhat arbitrary—any convenient juncture will work.

Notice that we've tacitly assumed time will go on forever into the future, but we could make this axis finite, that is, have a final point when time ends. After all, our universe isn't guaranteed to keep existing (Y2K or Nostradamus, anyone?). We also don't have to make time continuous—there are quantum theories where it exists only in tiny discrete instants. More exotically, we could conceive of a universe where time was able to run backward if it wanted, or even loop around in an endless cycle, so everything kept getting repeated (like in Hindu philosophy, or the movie *Groundhog Day*).

So many alternatives, so little time to explore them all!

Moving through spacetime

For now, let's stick with the idea of a single axis starting at 0. Now think of everything in the universe frozen exactly in its current state. Then the simplest way we could view time is that it strings together a progression of such states, one for each instant. (This is the way Newton conceived it.)

This "spacetime" presents an immediate representational difficulty. Since we already have three space dimensions, the additional time axis carries us into a four-dimensional realm—which we saw earlier is very hard to visualize.* That's why we've cheated in the picture above, by showing a globe instead of the full 3-D universe (a single copy of which would have overrun everything).

Let's therefore think of space being only one-dimensional for now, instead of three-dimensional. This means things can move back and forth only along a single axis, the *x* axis. Then, stacking copies of this axis, one for each time instant *t*, gives a simplified 2-D spacetime that's easier to represent.

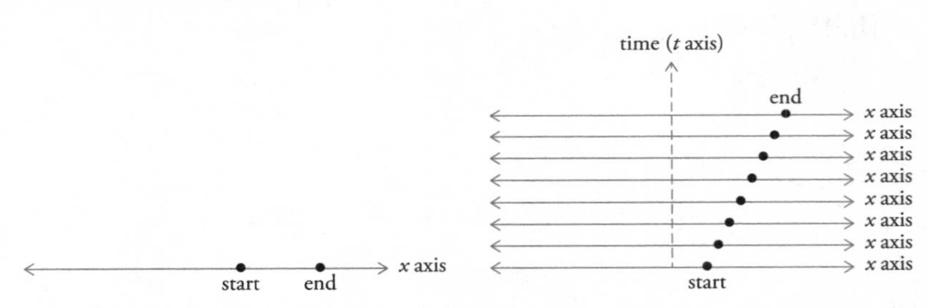

Consider now a particle that moves from one point to another along the *x* axis. Whereas previously we might have just represented this motion by showing the particle's start and end points on the *x* axis, we can now display its position at intermediate instants as well, like successive frames of a movie (see the diagram on the right above). More significantly, the concept of time allows us to measure not just *where* the particle moves, but how *long* it takes to move—we can read this off the *t* axis. In fact, let's define the concept,

$$\text{speed} = \frac{\text{distance traveled}}{\text{time taken}}$$

which you first saw years ago in school. This gives us a way to quan-

* Note that this is similar to the construction we used to create 4-D space, with time replacing the fourth space dimension now.

tify how "fast" the particle moves, and also compare its motion to that of other particles.

In the above picture, we showed only a few copies of the x axis, which correspond to a few different values of t. If we fill in the x axes for all the intermediate values of t, we end up with part of an (x,t) plane (this is similar to how we used the "matchstick stacking" method to construct our (x,y) plane earlier). Then, connecting all the different positions of the particle (one for each instant t) gives us a curve that represents the trajectory of the particle through our spacetime.

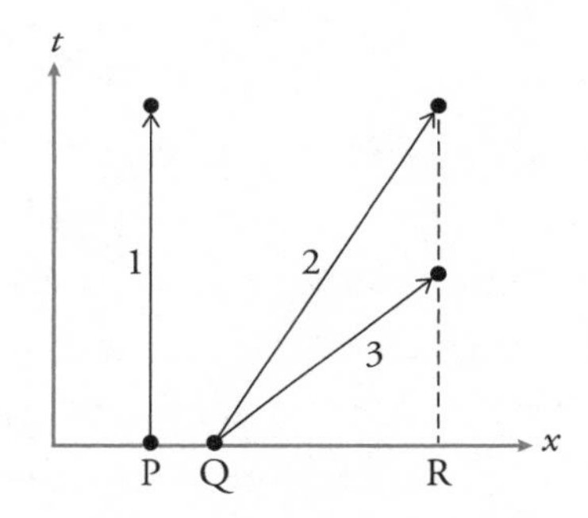

Above, we have drawn three such trajectories, corresponding to particles 1, 2, and 3. Let us interpret them. Particle 1 starts at point P on the x axis and remains there without moving. This is what a vertical straight line through our spacetime denotes (time t increases, but position x doesn't change). Particles 2 and 3 both start at the same point Q and reach the same point R on the x axis. The difference is that particle 3 takes less time, which means it travels at a faster speed. The closer the trajectory is to the horizontal, the faster the particle moves.*

The three paths through spacetime above are all straight lines.

* A perfectly horizontal trajectory would represent instantaneous motion from one point to another—a staple of sci-fi shows, but not something we actually experience in real life.

Could particles have trajectories that are curves? If so, what does such "curved" motion signify that is different from "straight line" motion?

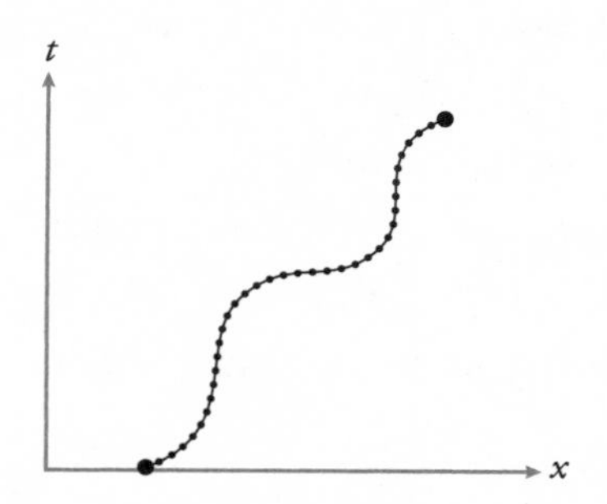

For the first question, the answer is clearly yes. Just plot a bunch of points in spacetime that don't lie in a straight line, and voilà! Connect them and you've created your own hypothetical particle with a curved trajectory, as in the picture above.*

Let's now see what such curved paths signify. Suppose we have two particles, both starting at A and ending at C at precisely the same instants. One follows a straight line, while the other's trajectory is curved as shown. What's the difference?

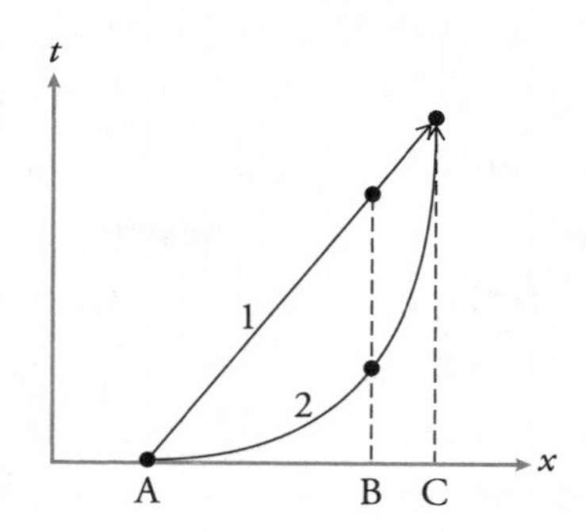

Consider particle 1 first. If B is *any* intermediate point, then the time this particle takes to reach B is always proportional to the distance

---

* Trajectories that place the particle at more than one spot at the same instant (e.g., the perfectly horizontal one) would again be disallowed.

AB. In other words, the motion is completely even, and the speed is exactly the same no matter which segment we're examining.

Particle 2 behaves very differently. It seems to zip along rapidly in the beginning, taking a much shorter time getting from A to B than particle 1 does. But then it loses steam between B and C, getting to C at the same instant as particle 1. So its speed is quite variable—fast between A and B, and then slow between B and C.

What we've called "speed" here, that is, distance traveled divided by time taken, is, more correctly, the "average speed" over that interval. If we take the interval to be AC, then this average speed is identical for the two particles, since they both take the same amount of time to go from A to C. The difference arises when we consider other intervals.

For particles whose trajectory is a straight line (like particle 1), the average speed is the same no matter which interval we measure it on. This is what characterizes straight-line motion through spacetime. Conversely, a curved path corresponds to this average speed being different over some different intervals. In other words, the distinction between straight and curved lines in our spacetime is just the distinction between constant and variable speeds.*

Straight or curved?

So far, Nature's simulated particles have had wills of their own. That's why they've been buzzing around in such undisciplined paths, dive-bombing us with impunity. However, to institute order in the universe, we're going to have to remove such free will—even though we may be unclear on how this might be accomplished (is it like

* In calculus, we also define an "instantaneous" speed at any point by looking at the limit of what happens to the distance divided by time ratio when the time interval is shrunk to 0. For straight-line paths, this instantaneous speed is again the same everywhere, whereas for curved paths, it will vary.

deveining shrimp?). Say we succeed; what law would we like to impose on particle motion? Should we always make particles go straight, or insist they take some curved path? For instance, should they always be made to travel in the arc of a circle?

Symmetry makes the decision for us. Recall that our 3-D space is "translationally symmetric." This means we can shift (i.e., "translate") our universe's space in the x, y, or z direction by a fixed amount and nothing changes. This is another way of saying that all of space is perfectly homogeneous, every cube of it is identical to every other. Consequently, if the universe consisted of a single particle in empty space, this particle wouldn't encounter anything that would change its motion. There will be no variation in the properties of the underlying universe—therefore, in the absence of free will, the particle's speed should remain the same. If it's at rest, it should remain at rest; if it's moving, it should do so at a constant speed, in a straight line. In other words, it should have a straight-line trajectory through spacetime. This is going to be our default.

Of course, there will be multiple particles in our universe, and these may attract one another, or even collide. That's going to require different rules, one of which we will consider in the next chapter. But the overarching principle, that symmetry leads to trajectories being straight lines, will continue to operate in the background. The uniformity of space also means that in the absence of other influences, laws should be the same everywhere in the universe, not differ from place to place.

Let's note that translational symmetry isn't the only symmetry our universe has. Its space is also "rotationally symmetric," that is, if we rotate empty 3-D space about any axis, it remains the same. Moreover, if we look at *spacetime*, rather than just space, then this is translationally symmetric as far as time is concerned, since the entire expanse of 3-D space is copied from instant to instant, and there is no difference between these copies.

Each of these symmetries plays a role in shaping laws—for instance, constancy with respect to time means that the same laws should hold tomorrow that do today. We won't go deeper into the exact laws that are shaped by the constraints symmetry imposes, but it's something to keep in mind: Nature's love of symmetry is intimately tied up with the kinds of laws our universe can have.*

---

* The profound connection between symmetry and Nature's laws was first expounded by the brilliant and insufficiently acclaimed mathematician Emmy Noether.

29.

THE CURVATURE OF SPACETIME

NATURE IS PLEASED WITH THE IDEA OF PARTICLES traveling in straight lines through spacetime. "As you know, I've always adored geometry," she declares. "Especially the perfection of the straight line." You decide not to bring up the question of why her lines are always so fuzzy if she's so enamored of straightness. The important thing is she has an intuitive feel for straightness, which will help you put into place a fundamental tenet regarding motion.

But there's a problem. You've just recently formulated the law of "gravitation," which is what you've named the effect by which particles attract each other. The goal is to have them agglomerate due to it, forming a larger body, with greater mass. As seen from the formula you derived earlier, this attraction is proportional to the product of the masses of the two bodies involved. What you expect is that the attraction between a small body and a much larger one may not affect the larger one too much, but will likely pull the smaller one out of its straight-line path.

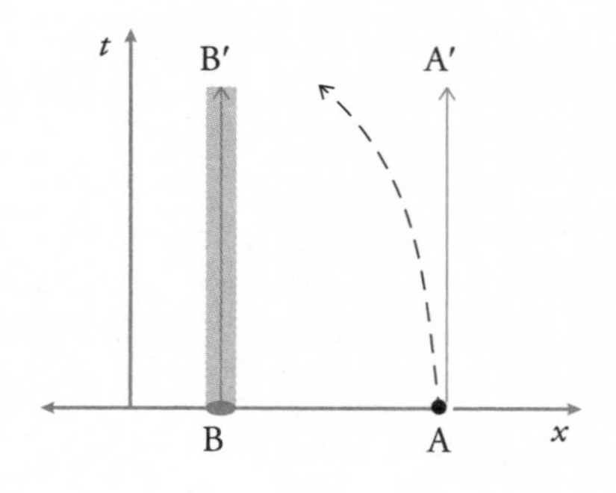

Take the particle A, which being at rest might ordinarily follow the vertical path AA'. Suppose, though, that there's a much larger body B at rest nearby, following path BB'. Then A's path might bend toward B's due to gravitational attraction. There might be some bending of B's path toward A as well, but we'll assume it's minor enough to ignore.

This bending leads to a conflict. We just told Nature that particles should follow straight-line paths. And here we are, already contradicting this, saying trajectories will likely need to be curved. Is there any way to reconcile these whiplash instructions, give our contractor just one rule to follow?

There is, and it's an elegant idea. Let me explain the gist. Assume the particles all try to follow straight-line paths, as we've told them to. When gravity exerts its pull, they do their darnedest to resist. They don't quite succeed, but due to their resistance, the curved trajectories that result are still the shortest achievable in the presence of gravitation's effects. In other words, these are the best possible alternatives to straight lines—a version of the shortest-path "geodesics" we encountered on Day 2, when different geometries were formed.

Recall what the term "geodesic" meant. When the underlying geometry was curved, as on the surface of a sphere, there were no "straight lines" in the usual sense of the term. Rather, we had to reinterpret such lines as great circles, that is, circles on the sphere with maximum radius (like circles of longitude).* The name "geodesics" signified that the shortest distance between two points always lay along one of these great circles (recall that's why airplane flight paths may go over the Arctic Circle). On a flat plane, these shortest-path geodesics were just the usual straight lines, while on a hyperbolic plane, they had yet another (curved) interpretation, which we pictorially described.

The situation now is very similar—except that instead of geometry driving the substitution of "geodesic" for "straight line," it's gravitation. What if we linked these two causes by designing a universe in which gravitation actually made the geometry change? Then straight

* See the illustration on page 110.

lines would bend, just like on the surface of a sphere or hyperbolic plane. A particle trying to go straight would be forced into a geodesic curve by gravitation, but this effect would be indirect—gravitation causing the underlying geometry to change, and this in turn causing the particle's straight-line path to bend.

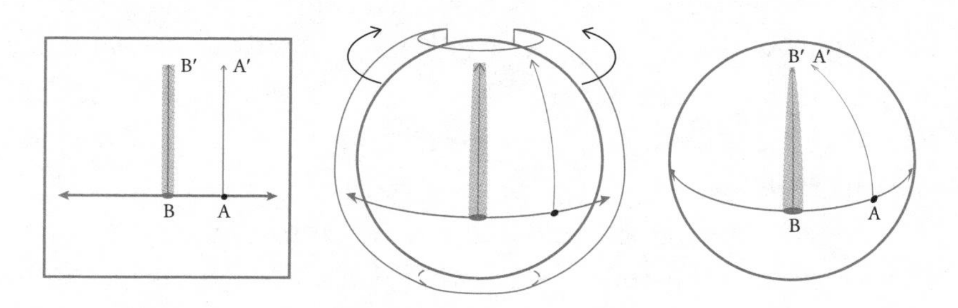

As an illustration, consider again the particle proceeding along AA'. Suppose gravitation were to somehow warp the underlying plane to wrap around a sphere. The simplest way to accomplish this is by the same method used to project a rectangular map of the world onto a globe*—let's assume this is what's used here. Then the particle has no choice but to be pulled toward the larger body. Indeed, the parallel "shortest-distance" paths for A and B will now align with the sphere's geodesics—and converge. So introducing such curvature into our 2-D spacetime makes A' close in on B', just as gravitation demands. Best of all, the linear-path rule still holds—as long as "straight lines" are interpreted to mean "geodesics." Our new one-size-fits-all directive to Nature would simply be "Particles should move through spacetime along geodesics." (Here, spacetime could be curved or not.)

The above spherical warping doesn't quite give us what we want. We'd have liked the bigger body to have more gravitational pull than the smaller one, but that's not possible here. Put any two particles at

* The mathematical formula this involves, called the "equirectangular projection," has been known for almost two millennia. When this formula is used, lines of longitude on the sphere become vertical lines on the rectangle, and vice versa. (Lines of latitude are similarly mapped onto horizontal lines on the rectangle.)

any two points on the sphere, and they'll always converge, exactly the same way—it doesn't matter what their masses are. In other words, *all* particles at rest eventually get pulled to one another, so what we have is a rather simplistic gravitational model.

But one that does convey the gist of the idea. The sphere is familiar to all of us—it's the non-Euclidean geometry we're most comfortable with, which is why I've used it here. For a more discriminating model, the geometry would need to bend only locally, and proportionally to mass, in the vicinity of where that mass was situated. Popular diagrams, like the one below, help visualize this. But treat such pictures as metaphor, not accurate representations, since while the objects are 3-D, the space they're curving is only a 2-D plane. Moreover, time seems to have been entirely omitted.

For our universe under construction, it wouldn't be just a plane being deformed but full four-dimensional spacetime. This presents an additional problem. Just as we needed the 2-D plane to be embedded in three-dimensional surroundings to help us picture how it might bend or deform, so also we'd want this 4-D spacetime to be embedded in *five* dimensions to best picture how it might develop curvature. There's no way we can visualize such 4-D or 5-D. It's the problem of the blind men and the elephant again.

But the point is not to visually flesh out the elephant. Rather, it's to be able to mentally absorb two related ideas, for which the preceding diagrams should provide sufficient aid. The first idea is that when you curve the underlying geometry, straight-line paths (along which the distance traveled is minimized) get replaced by curved geodesics. The second is that this allows bodies to exert gravitational attraction—by curving spacetime with their mass, they make the resulting geodesic paths veer toward them.

Which brings us to the promised reference to Einstein. The above two ideas were fundamental ingredients in helping him incorporate gravity into his general theory of relativity. The actual theory (on which there are excellent popular works written by physicists) is too complicated to include here (see endnotes)—so we're going to have to be content with our bargain-basement gravitational model. What general relativity posits, for instance, is that the sun's mass curves spacetime in the vicinity of our solar system. This is why planets revolve around the sun rather than flying off in straight lines—they are just following the curved geodesics in this deformed spacetime.

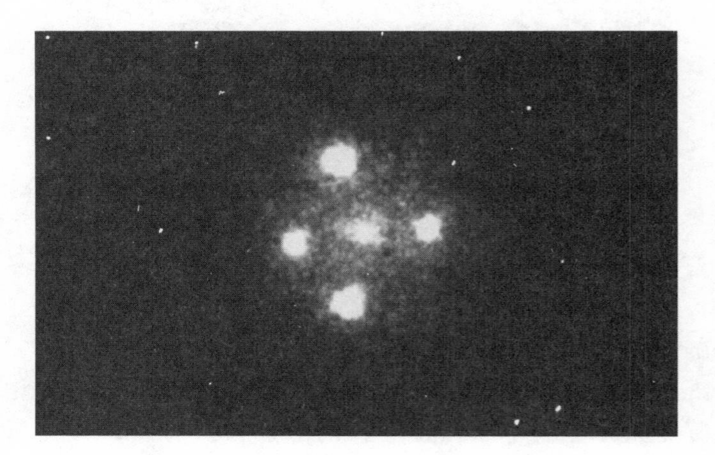

Is this really what happens, though? Some of the most compelling evidence to show that spacetime is truly curved in our physical universe comes from observing light from distant astronomical sources get distorted by massive galaxies. For instance, the photo above, of the

"Einstein Cross," shows four images of a quasar* situated completely *behind* one such galaxy. By all rights, the quasar should have been completely blocked by the galaxy. Instead, the four images indicate that light rays traveling from the quasar that ordinarily would have been directed straight in other directions (and lost to space) are bent by the galaxy's gravity to become visible to us. However, light—in the form of photons—is massless, so shouldn't be affected by gravity. The only explanation, therefore, is that the galaxy has actually curved spacetime locally, and this is why the light rays (following geodesics now, instead of straight lines) reach us.

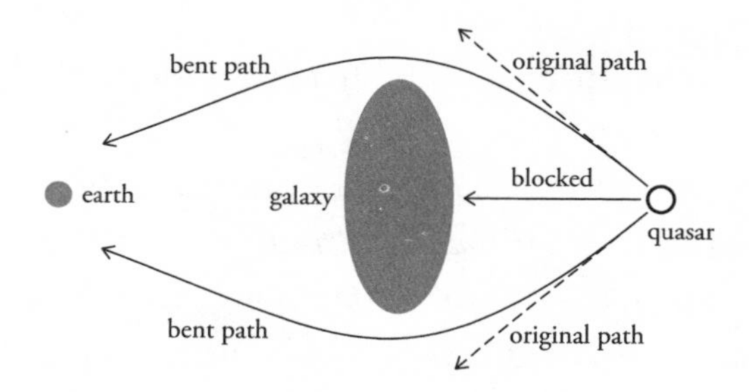

We've been building up to this moment ever since Day 2, when we first questioned whether the line segment between our first two points had to be straight. Developing curved geometry has taken us out of our comfort zone, made us question fundamental assumptions—much like the nineteenth-century mathematicians who first discovered there could be consistent geometries besides the flat one proposed by Euclid. You may have wondered why we were pursuing something so esoteric—couldn't we have just stuck to flat geometry to simplify our book's narrative? This, finally, is our payoff. Non-Euclidean geometry isn't just a theoretical exercise; it can play a fundamental role in any universe we might build, just as Einstein found it did for the one we live in.

* Quasars are faraway cosmic objects of very high brightness, found near the centers of some galaxies.

The question physics asks of us

With our new understanding of space and time, some previous questions about the universe's extent get relevant again. We've assumed that although spacetime can curve in localized spots* due to gravity, the vast majority of it is flat—which suggests it would be infinite. If, however, it curved in on itself like a sphere or doughnut, then spacetime would be finite. We aren't bringing up these possibilities here to delve into them further but, rather, to remind us we've always used the term "infinite" rather cavalierly. Before applying it to our universe, shouldn't we understand better what infinity really means?

It's not just the *extent* of space and time where the infinite comes into play, but also their *continuity*. Infinity is what allows us to proceed along an unbroken path of points along any trajectory, whether it is a continuum of coordinates in space or a stream of successive instants in time. Can one enumerate such points in a trajectory, say from the first to the infinitieth? What is the nature of the infinity that characterizes such a continuum, exactly?

Infinity also crops up in the context of whether to impose any restrictions on the *size* of the quantities we introduce in our universe. For instance, we've talked about mass and speed—can these be infinite? Even if we decide they cannot, can they get arbitrarily large? Or should we put a maximum bound on them, which they

---

* A galaxy and its environs might hardly qualify as a "localized spot," but remember, the universe is almost completely empty, with all matter occupying only a minuscule blip of its total space (0.0000000000000000000042 percent, by one estimate).

cannot exceed? Such decisions will have fundamental repercussions for our universe.*

Physics isn't the only subject that prods us to probe this as-yet-hazy aspect of numbers. Most religions also seem to subsume the infinite. How can we talk of "omniscience" without knowing how far the body of mathematical knowledge might stretch, or reconcile "eternity" with the endlessness of time? Surely the pope would appreciate any additional light that mathematics might shed on such theological concepts.

So let's explore this basic idea—infinity—that's been an essential thread running through the design of our universe. Is infinity a number, or is it something else? What are its mathematical properties? We need to settle such questions to bring completion to our thought experiment. Our ruminations on infinity will lead us further than we might anticipate, raising some fundamental philosophical questions into the nature of knowledge itself.

---

* Indeed, Einstein's special theory of relativity posits a maximum speed in the universe: nothing can travel faster than c, the speed of light. This has various consequences. For instance, a particle's mass has to start approaching infinity when the particle's speed approaches c, since otherwise, there would be nothing stopping the speed from exceeding c if the particle was given more and more energy to accelerate it (time also slows down at speeds near c for the same reason). So if we try to bound one quantity (e.g., speed), it might force another (e.g., mass) to become unbounded.

Day 6

INFINITY

Exploring the role of infinity in the universe and beyond

30.

A FINITE UNIVERSE OF NUMBERS

NOW THAT I'VE GOT YOU ALL GOOD AND READY TO learn about the infinite, let me throw you a curve ball (a googly, I believe they call it in cricket). What if we don't really need infinity? What if the universe we design can get by just fine on the finite, thank you very much?

The possibility would fly in the face of everything we've done. After all, our universe started with the naturals, which we know continue without end. Surely a universe based on numbers should be infinite like them.

But what if we've overbuilt our number system? Just because we've indulged our intellects to create all these constructs doesn't mean we're obligated to use them. Why not just extract the minimum set of numbers needed to make things work? A *finite* set, it might turn out.

Absurd? Nonsensical? Before you dismiss this idea, let me assure you that such finite universes of numbers have already proved their mettle: most computers operate on them.

A system with a smallest and largest number

Computers are machines—they have finite memory, finite computational power. While in principle we can represent numbers up to billions of digits in the memory of current computers, in practice, widely used software will cut off a number such as 1/3 after a small number of digits (such as 16 or 32) because computation becomes very efficient with such representations. Therefore, you won't get an infinite number of 3s, as in the expansion 0.3333 . . . . Similarly, pi (or any other irrational) will also be truncated after the same number of digits. As we mentioned earlier, approximating pi this way does not cause practical problems for NASA. They have been truncating it after 15 decimal digits, which turn out to be sufficient for even the most delicate calculations.

There's more—computers generally have a maximum number M they recognize, beyond which any quantity is marked as "overflow" or "infinity." The same goes for a smallest decimal d, say of the form $d = 0.00 \ldots 01$, where the number of zeros may be 40, 100, 400, whatever. Anything smaller, and the computer just tags it as 0.

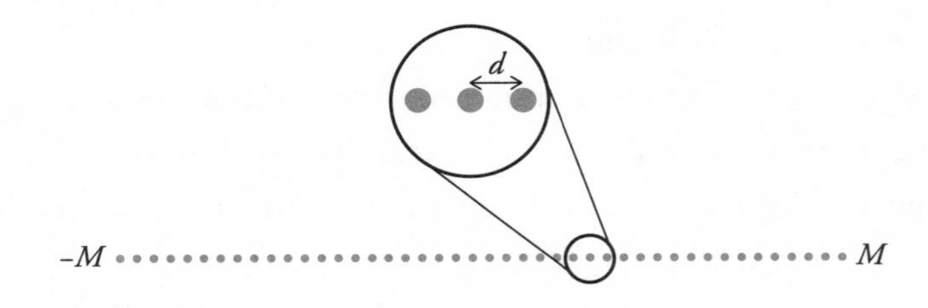

In its simplest form, such a computer system therefore consists of a finite grid of numbers, with some uniform spacing d. The grid goes from $-M$ to M, so that the set of reals is not only hole-ridden, but truncated. For numbers like 1/3 or pi that aren't included, the computer substitutes the nearest approximation for the exact value. Anything more than M, it treats like infinity.*

* What we've described is a so-called "fixed point" number system, where the grid of numbers is uniform and always separated by the same fixed spacing d. Computers generally use "floating point" systems, which, though again finite, are more sophisticated. In these, the grid is much finer close to 0 and gets sparser as you progress to larger numbers; i.e., the spacing d varies.

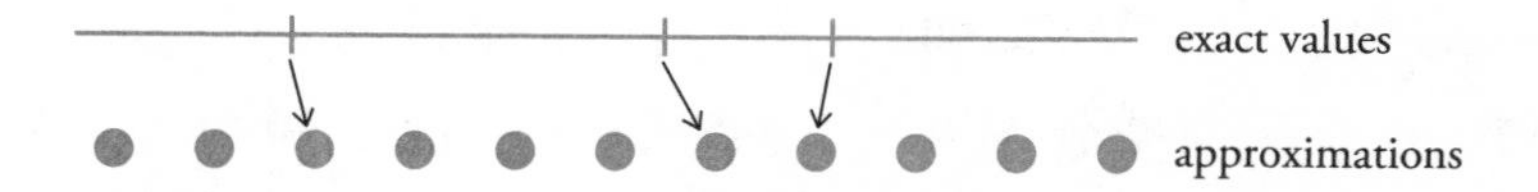

As long as M is taken to be sufficiently large and d sufficiently small, the loss in accuracy only rarely causes a problem, especially once additional safeguards are built in.

A universe of grid points

So how does this idea translate to the universe we're building? The crux would be this: instead of continuous and infinite, what if the universe were discrete and finite? What if space consisted of unconnected nodes, arranged in a grid? In such a scenario, each axis would go from $-M$ to M. That's all there would be—the missing points would simply not exist. The grid in the illustration shows a section of such space.

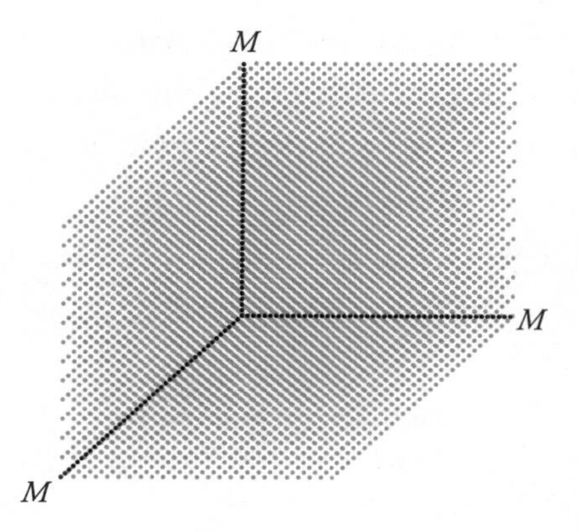

Does this sound implausible? If so, let me take you back again to when we were creating geometry, and the first line emerged to connect two points. Earlier, we asked whether the line was straight or curved. Now we'll ask whether the line had to necessarily be a continuum. Why couldn't only a finite number of intervening points have emerged instead, with the ends of the line extending only until they reached some finite limit? Recall that one of the main functions of the line was to provide condo units for all the reals. If we took our underlying inspiration to be a finite set of numbers, then a finite line, with a finite number of condo units, would be sufficient.

Movement in such space could still occur, but it would be from point to point, much like a pulse running through a string of Christmas lights. One could ask how a particle might "jump" across the intervening gaps, but this is not a valid question, because such gaps wouldn't exist. Since we start with nothing, the grid of points would be all the "space" our universe contained. The same principle would hold for time, which could now be experienced only in discrete instants. Gaps in between would again not exist.

Certainly, we can make a go at formulating our universe-in-progress based on such discrete space and time.* In doing so, though, we would need to rethink many of the concepts we've been developing. For instance, would particles now have to exist localized at a single grid point? Would they still have volume, and if so, how would we define it? Perhaps we could make each point be a tiny sphere instead of a dimensionless geometric point, to resolve such issues. But then, inside each sphere, we'd be back to having a continuum of space—the very concept we were trying to avoid.†

Whatever the particulars, having a discrete universe would mean our understanding of both length and time would have to change.

---

* In fact, models involving both discrete space and discrete time are used in quantum theories for our own universe. Further, physicists may not agree on whether the reality of space and time in our universe is a continuum or has such discrete quantum character.

† Some physics theories use the concept of a minimum length in the universe (equal to about 10^{-35} meters and related to the so-called Planck's constant), such that at levels below this minimum, space becomes granular rather than continuous, i.e., you reach the discrete building blocks (or quanta) of space. A similar "smallest meaningful interval" (of about 10^{-43} seconds) is theorized for time.

We'd have to scrap much of the geometry and physics we've developed in this thought experiment and start from square one again. There's no good reason to do so, given how sound our constructs feel, how naturally they spring from the numbers we've created. Let's keep the continuous model we've built.

Besides, ideas, once born, cannot be erased. Even if we were to backtrack, the concepts we've developed here would remain. We'd still have our number systems, our geometry, our design for the universe; there'd still be open questions about the meaning of infinity and the mysteries of the continuum.

This, then, is perhaps the ultimate reason we must go forth to explore infinity. A computer might be content to remain confined between $-M$ and M. But we've already caught a glimpse beyond. Our need to know as curiosity-filled humans urges us on.

31.

CLOSE ENCOUNTERS OF THE INFINITE KIND

A CENTRAL TENSION THAT DRIVES MATHEMATICS IS the push and pull between the finite and the infinite. Think of how we started, using nothingness to create 0 and 1. This gave us a process we could keep repeating: 1 begetting 2, which begat 3, and so on. That's the first inkling we got of infinity—a name for this endlessness, which, like an invisible moon generating tides, keeps pulling the process along. In doing so, it draws 1, the basic unit of finiteness, through the series of natural numbers 2, 3, 4, . . . , toward the unreachable realm of the infinite.

We encountered such endlessness repeatedly in our explorations. Each time we wanted to specify an irrational number like $\sqrt{2}$ or pi in decimal form, we were faced with the task of writing down an endless string of digits. The same held for fractions like $\frac{1}{3}$ or $\frac{2}{7}$; in fact, for most real or complex numbers. In other words, these number systems couldn't exist without invoking infinity, in the form of endlessness, in their construction.

Geometry depended on this notion of infinity too, right from the very first line segment that extended endlessly in each direction to form a straight line. Planes were formed by joining together an endless series of straight lines, and space by laminating endless layers of planes. Fractals needed a process to be repeated endlessly as well, without which they would lose their "fractalness." For instance, the Sierpinski triangle fractal would be just an ordinary geometric pattern of a finite number of triangles if only a few steps were carried out with the input/output process.

Physics, as we just saw in the previous section, also depends on endlessness. Otherwise, not just space but also time would terminate at some point. For all we know, spacetime is still forming, expanding, much like the never-ending generation of the naturals. What this would imply is that, while space and time are finite at any instant lived by us, if this endless expansion were ever completed, their limit would be infinite.

Of course, we *can't* complete any of the above endless processes—at least not physically. We cannot list all the naturals or exactly write down pi, we cannot draw a complete straight line or realize a fractal, we cannot ride an expanding wave of spacetime to infinity. And yet, we can conceive of all of these, and in this sense, bring everything to fruition. Starting with the Big Bang of Numbers, which gives us the complete set of naturals.

Such conceptual constructs give infinity a second, enhanced meaning. A process that can be continued indefinitely carries only the *potential* for infiniteness. That's because we know that no matter how many repetitions we perform, we'll always end up with something finite. However, once we conceive of the entire set of naturals, or an entire straight line, or a fully formed fractal, we've mentally realized this potential to get an *actual* infinity. We've overcome the barrier of endlessness to arrive at a deeper concept, a view to a completed limit beyond. That's why the Big Bang of Numbers, even if it's only in our minds, is so momentous, so transformational.

The godfather of numbers

So how should we picture infinity? Should we take synesthetic license and imagine it as an entity located at the end of the list of naturals? Given the crucial role infinity plays in birthing naturals (the source from which everything in our universe arises), perhaps a Godlike persona would be appropriate. In this context, physics also uses infinity in creation, by having the big bang arise from a concentrated source of infinite density.

But infinity doesn't create the naturals by itself—Zero or One should surely get equal billing. So the "God" appellation isn't quite appropriate. And yet, infinity is always lurking around, using hidden strings to shape much of mathematics. Perhaps the way to think of it, then, is as an invisible puppeteer, a godfather in the Mafia sense of the word. Someone always stirring things up, accentuating the tension with the finite.*

Whether God, a godfather, or just a marker denoted by the catch-all symbol ∞, it's useful to picture infinity situated at the end of the naturals or the reals, to signify their infinite limit. The same idea applies in geometry, to any straight line. Imagine a point P move along the line, then its distance OP from the origin keeps increasing as P travels in either direction. The fact that this distance gets boundlessly large means the ends of every line tend to the infinite.

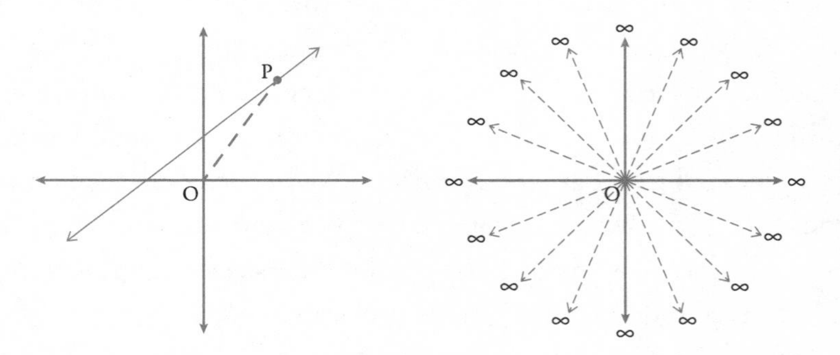

We can therefore picture infinity, in this sense of distance, not just at the ends of any line but also at the edges of any plane. Since distance

* In case you're wondering, "The Godfather of Numbers" was the working title for an earlier draft of this book. The idea was to emphasize how pervasively infinity influences all of mathematics.

is also defined in higher dimensions, we can picture ∞ at the limits of 3-D space, not to mention 4-D and 5-D as well.

From the above, we see that infinity can be interpreted as a quantity greater than any natural or real.* But is it a number itself? It obviously can't be a real number or a natural, since it's greater than any of them. Neither can it be a complex number, since it lies just beyond the extremities of the complex plane. One could think of it as an "extended" number of sorts, but this is a matter of semantics. What infinity is, like any other number, is a concept.

Does this concept *really* get realized in the physical universe we're building? We've seen all the cases—from fractals in clouds and shells to the endlessness of time—where it remains only a potential infinity. Which is fortunate, since Nature, as we've noticed with her simulated particles, seems so far capable of working only with the finite.

And yet, there's one striking exception—the continuity of space and time. Infinity is inescapable here, unless we're willing to jettison our universe and create a new discrete design. We know that between any two reals, there exists an infinitude of other reals—a basic requirement for the condos we created on the real line. The same holds true for the instants between any two points in time. So our picture is not just of infinity roosting at the unreachable extremities of planes and such, but also embedded in every jot of space, each bit of every timeline.

This brings up the practical question we posed earlier. To go from

---

* This characterization finally allows us to understand what happens if we try to divide 1 by 0. We know $1 \div 0.1 = 10$, $1 \div 0.01 = 100$, $1 \div 0.001 = 1000$, and so on. If we keep dividing 1 by smaller and smaller numbers, the quotient keeps increasing. In the limit, we'd get an indefinitely large value, i.e., one greater than any real. This ought to equal infinity. The problem is that we could also divide 1 by $-0.1, -0.01, -0.001, \ldots$ to get $-10, -100, -1000 \ldots$, i.e., values that are increasingly negative. As the denominator approaches 0, we would now get *minus* infinity. So we can't ascribe a unique infinite value to $1 \div 0$ (it could be $+\infty$ or $-\infty$). We therefore say it's undefined. For the same reason, $x \div 0$, for any real $x \neq 0$, is also undefined. (We already saw in Chapter 3 why $0 \div 0$ is undefined.)

point A to point B, we have to pass through the infinity of points in between; the same holds for passing from one instant to another in time. Can we number these intervening points or instants? Count them, in other words, as the first, the tenth, the fortieth, and so on? After all, the infinity we started with—that of the naturals—certainly allows such numbering. Also, we're traversing these points in progression, so surely it should be possible to numerically express this ordering, right?

To answer this question, we need to look at infinity in another way—one that probes deeper into its meaning. We need to understand it in terms of a concept called *cardinality*.

32.

THE MATCHING GAME

THE POPE CAN'T HELP SMILING AT THE WORD "CARdinality." He remembers the moniker, "His Royal Cardinality," they used to give cardinals who were unduly bedazzled by the trappings of their position. He imagines a gaggle of them, red-robed and hunched over a table, trying to puzzle out the word's mathematical meaning.

Something about infinity jogs his memory—a reference in an account of Pope Leo XIII he's been reading. His nineteenth-century predecessor's message was simple: the Church didn't have to be at loggerheads with science, they could work with each other, compatibly. History provided several such examples: the monks in the Dark Ages who kept the last vestiges of science and mathematics alive in the West, the Catholic cathedrals that became medieval centers of learning and gave rise to the first European universities. The pope is eager to promote Leo's message in the current era. With climate change escalating and lethal viruses running amok, he knows how indispensable science is going to be.

The infinity reference was no more than a footnote, really. Something about a mathematician sending Leo pamphlets to try to explain infinity. It sounded so quirky and wonderful—this meeting of mathematics and religion—that the notion still lingers in the pope's memory. Were the materials preserved? Could they still be in the Vatican library? What was the mathematician's name—Georg something?

The pope turns back to the advance copy of my book. Apparently,

Leo wasn't the only pope to be targeted by unsolicited missives from a mathematician, he muses. This "cardinality" has him intrigued, though. Perhaps he'll read one more chapter before going to sleep.

The hand with infinite fingers

I hope the pope won't be too disappointed when I disclose that cardinality is not about ecclesiastical rank, but counting. Which is why I'm going to suggest we use hands and fingers to understand it (even though we technically haven't created these yet, and it's not clear how we're going to, in the book's remaining few chapters).

Here's a key point about these "digital" tools. Even if your teachers had never taught you about numbers, you would still have been able to tell that one hand has exactly as many digits as the other. All you'd need to do is establish a matching between fingers, a *correspondence* as we'll call it.

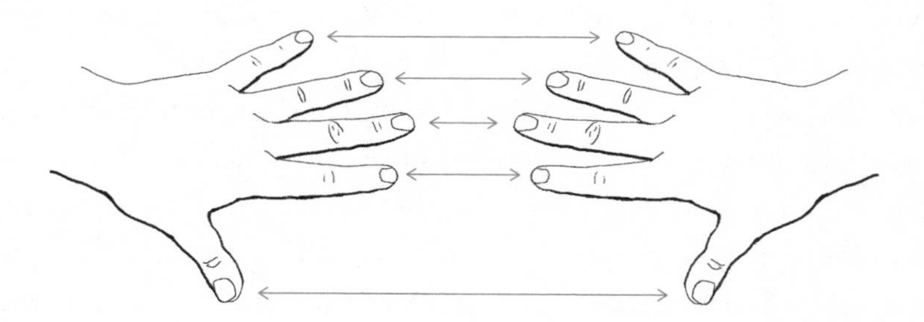

Suppose you came across a creature—say, a bird—and wanted to see if its digits were equal in number to yours. You could play the correspondence game again. This time, though, let's say there wasn't a perfect match. Depending on who had digits left over, you'd conclude you had either more or fewer digits. For instance, with birds, which generally have four digits (as do Homer Simpson and most other cartoon characters, since it makes the drawing easier), you'd come out ahead.

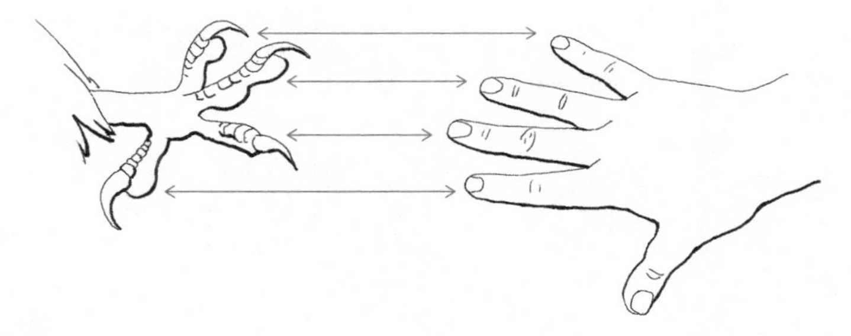

Notice you could play this game no matter what the number of fingers on each of your hands was—five, ten, a hundred, a million. In each case, if there was a perfect match, you'd conclude that the number of fingers on both hands was the same.

A hand with a million fingers is hard to imagine. But actually, we all have access to a phantom hand with an *infinite* number of fingers. It's called the set of naturals! When we first learn to count, we might establish a correspondence with a subset of the fingers of our hand—for instance, *three* apples or *four* oranges.

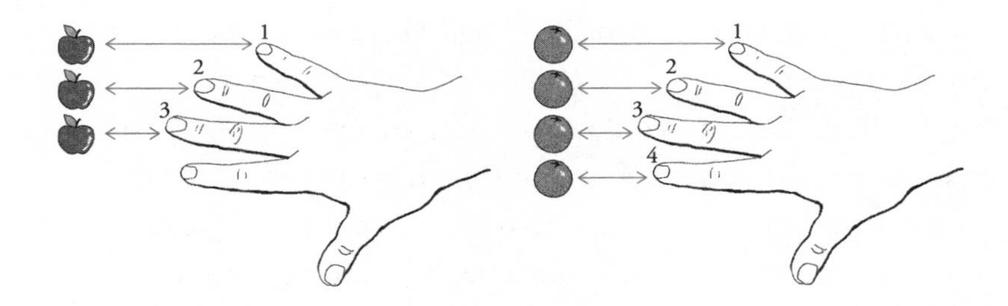

Assume we use only one hand, then we run out of fingers at five.* After this, we learn to use the phantom hand of naturals instead. Confronted with *seven* pears, for example, we establish a correspondence with the subset of the first seven naturals, that is, the first seven fingers of this invisible hand.

* The question of why we have five fingers instead of four or six remains unresolved from an evolutionary viewpoint. If humans had six fingers on each hand, we'd almost certainly be using 12 as our number base now.

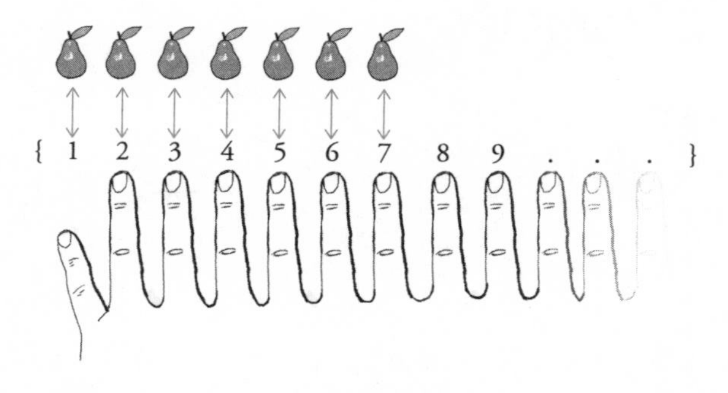

What if we now had an *infinite* set of pears? (Surely if we can imagine infinite fingers, we can imagine infinite pears.) Could we say something about counting them as well?

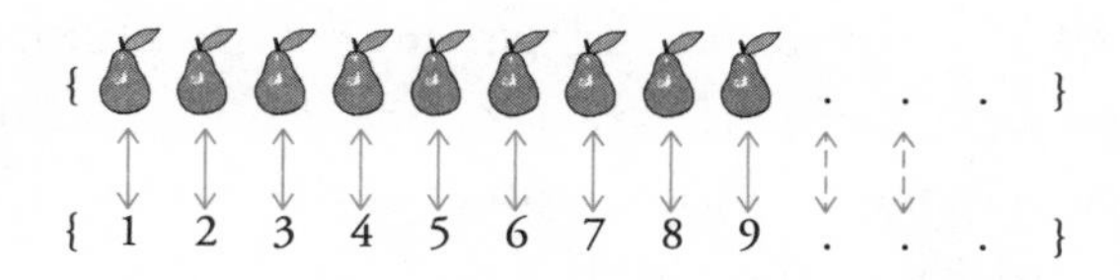

Why not? Suppose we were able to find exactly one pear in our list corresponding to each natural, and also exactly one natural corresponding to each pear—in other words, establish a one-to-one correspondence. Then we could reasonably conclude we had exactly as many pears as naturals. This kind of matching would allow us to assign a count to infinite sets. Or, to use a fancier term, a "cardinality."

The cardinality of a set of three pears is 3, as is the cardinality of a set of three numbers such as {1, 2, 3} or {5, 26, 11}. And when we examine our entire hand of naturals, then its cardinality is infinite, which we'll denote by our usual symbol, ∞, at least for now. This ∞ will also be the cardinality of any set—say of the infinite pears above—that can be matched one-to-one with the naturals. But can *all* infinite sets be matched with the naturals?

To better immerse you into the world of questions this raises (and also to experiment with how a little bit of fiction might help us understand math, I'll admit), let me phrase these issues in terms of a crucial ingredient our thought experiment has been lacking so far: interplanetary warfare.

Star wars

Imagine a galaxy far, far away, with two planets, Aleph and Cee, which are just like Earth, except for one peculiarity they share. Instead of just potential or conceptual infinity, they've realized *physical* infinity. So they have infinite material, infinite resources.

The Alephites are a peace-loving people, unlike the Ceenums, who have always had designs on their neighboring planet. One day, the Ceenums announce the deployment of a laser weapon that can blow up all of Aleph. "Surrender or be obliterated" is the text message millions of Alephites receive from the Ceenums on their interplanetary mobiles.

Fortunately, the Alephites have been preparing for such hostilities. Over the years, they've built a defensive laser of their own—one that can be aimed at the Ceenums' weapon and completely neutralize it. They wheel it into place at once and text their aggressors back that any attack will fail.

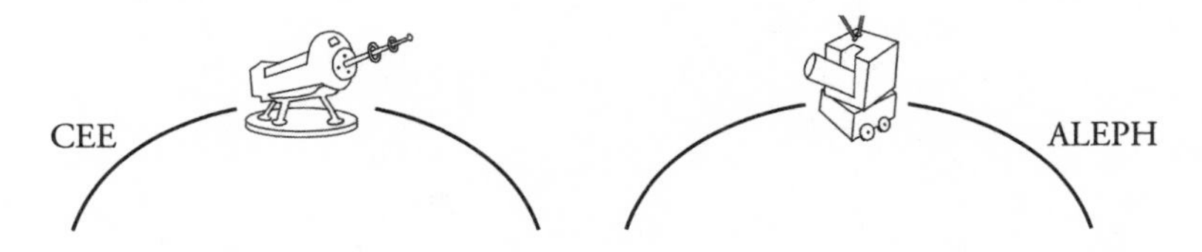

Rather than develop a new weapon, which could take decades, the Ceenums simply build a second laser. After all, they constructed a factory for such production, and this is the easiest way to escalate their threat. But upon its deployment, the Alephites are quick to match it with a second defense laser of their own. The two sides are still tied in an even contest.

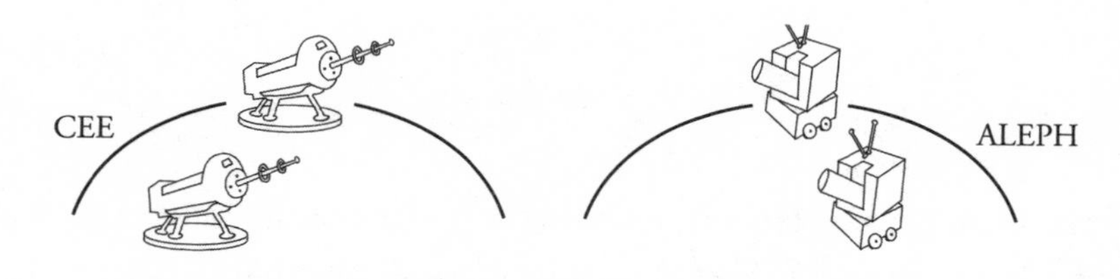

The situation intensifies rapidly after that. An arms race that escalates faster and further than anything between the Americans and the Russians. Two lasers each lead to three, then four, then five, and so on. And since infinity is just an everyday quantity on both planets, the two sides soon have as many weapons each as there are natural numbers.

At this point, a joint interplanetary peacekeeping effort is instituted. A team inspects each side's weapons and marks them with natural numbers 1, 2, 3, . . . , to ensure they're exactly matched. This clearly demonstrates that each side has a lock on the other. So it would be futile to start any war.

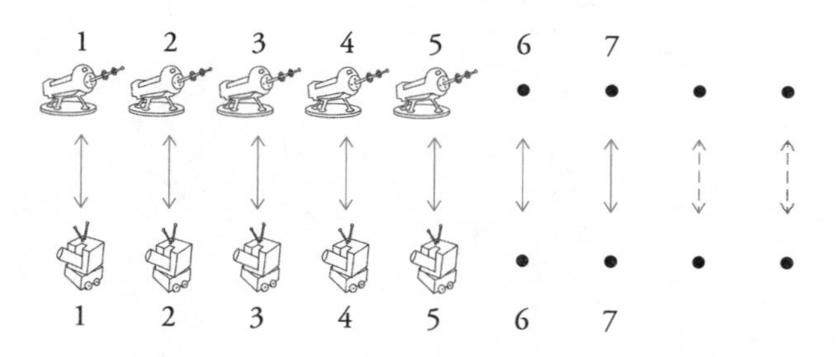

But unbeknownst to the Alephites, the Ceenums have slipped in some saboteurs among their peacekeeping representatives. While all eyes are on the numbering activity, these infiltrators manage to completely incapacitate the laser factory on Aleph. The damage is so thorough that the Alephites realize they'll need to start from scratch to build more weapons. They're stuck with the arsenal they have, at least for the next several years.

Needless to say, the Ceenums lose no time pressing their advantage. They create just one more laser, painting a 0 on it, in the same style the peacekeepers had numbered the rest of the arsenal. "It might be a zero," they text, "but it's enough to blow you all to smithereens."

Panic erupts on Aleph. With their factory sabotaged, there's no hope of matching planet Cee's "Laser 0." How tragic that after the infinite number of defenses created, a single extra weapon could be enough to vanquish them. The Alephites prepare to accept defeat rather than see their planet destroyed.

Just as their leaders are about to signal their surrender, Georg, one of their mathematicians, intervenes with a rescue plan. He refocuses their planet's Laser 1 on the Ceenums' new Laser 0. Then he trains the Alephites' second laser on the Ceenums' Laser 1. Proceeding this way down the line, he once again demonstrates a one-to-one correspondence, just as the peacekeeping force did. The two sets of weapons are in another evenly matched lockdown!

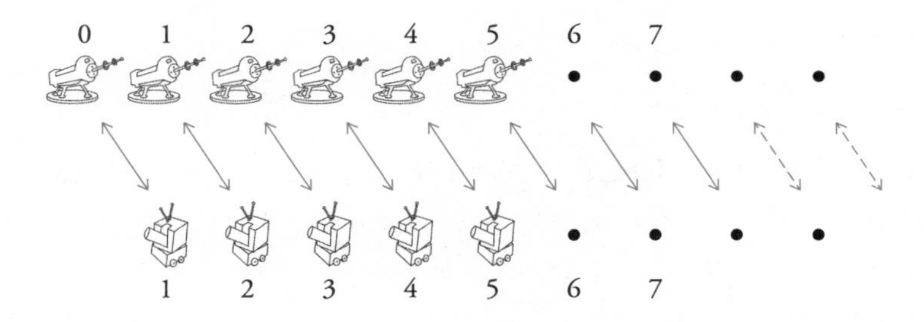

Not only that, but each time the Ceenums up the ante with another laser, Georg simply changes the starting point for the Aleph lasers. Since each arsenal is infinite, he is able to establish parity again. The Ceenums give up their strategy, and the Alephites declare Georg a hero. Thanks to him, they've beaten back what had appeared to be an invincible attack.

✳

Another threat

Some weeks after their celebration, though, a startling report comes from the lookouts at the telescopes. The Ceenums have regrouped, with an arsenal twice as large as before. They've constructed enough new lasers to have one for every integer*—positive, negative, and zero!

Clearly, Georg's strategy will be inadequate now. The problem is that while the naturals are only infinite in one direction, the integers stretch to the infinite along both the positives and the negatives. So if the Alephites matched their naturals to the Ceenums' positive integers, they'd have nothing left over to match the Ceenums' negatives with!

Ceenums	{ ... –3	–2	–1	0	1	2	3	4	... }
					↕	↕	↕	↕	
Alephites					{ 1	2	3	4	... }

Once again, it falls to Georg to rescue his compatriots from the brink of surrender. He realizes the trick is to alternate the positives with the negatives while locking in on the offending lasers with defense weapons. He mentally rearranges the Ceenum lasers using their numerical labels, then matches this rearranged list to his list of naturals. This has the effect of enumerating all the integers in a single list: 0 is the first integer on this list, 1 is the second integer, –1 is the third integer, and so on.

Ceenums	{ 0	1	–1	2	–2	3	–3	... }
	↕	↕	↕	↕	↕	↕	↕	
Alephites	{ 1	2	3	4	5	6	7	... }

The Ceenums are stunned when their hordes of new weapons are

* For some readers, this might be a good spot to take a peek at the table on page 63, which summarizes what we mean by integers, rationals, irrationals, reals, etc.

neutralized by the Alephites—who, rather than construct any new lasers, have just relied on rearrangement. They're forced to retreat, and peace once again reigns between the planets.

So the moral of the story is that the integers are exactly as numerous as the naturals, even though there seem to be about twice as many of them! We can use ∞ to denote the cardinality of both sets.

You might find this hard to accept. The set of integers is clearly larger, you might protest, since it contains the naturals as a *proper subset.**

Ah, but be careful. You're mixing two terms: "larger" and "higher cardinality." They mean the same only for sets that are finite. In fact, give me any infinite set and I can match it one-to-one with a "smaller" subset of itself! For instance, the naturals can be mapped onto the evens, which again constitute a proper subset. The same goes for the odds as well.

$$\begin{array}{cccccc} \{ & 1 & 2 & 3 & 4 & \dots \} \\ & \updownarrow & \updownarrow & \updownarrow & \updownarrow & \\ \{ & 2 & 4 & 6 & 8 & \dots \} \end{array} \qquad \begin{array}{cccccc} \{ & 1 & 2 & 3 & 4 & \dots \} \\ & \updownarrow & \updownarrow & \updownarrow & \updownarrow & \\ \{ & 1 & 3 & 5 & 7 & \dots \} \end{array}$$

But your dubiousness is well placed. The question you really want to ask is whether there's a set with cardinality greater than ∞. Maybe the set of fractions, that is, rationals, might be it? There seem to be so many more of them compared to the naturals. For instance, between 0 and 1 alone, there are already an infinite number: the fractions $\frac{1}{2}$, $\frac{1}{3}$, $\frac{1}{4}$, $\frac{1}{5}$, . . . , along with $\frac{2}{3}$ and $\frac{3}{4}$ and $\frac{2}{5}$ and $\frac{3}{5}$ and $\frac{4}{5}$ and so on. How would you ever count them? Arrange them as the first rational, the second, and so on?

You remember back when you were first experimenting with having numbers light up their points on the real line. When you instructed all the naturals to light up on the positive real axis, the

* Set B is a *proper subset* of set C if every element of B is in C but there is at least one element of C that is not in B. For example, let B be the set of ants and C be the set of insects, then B is a proper subset of C.

points of illumination were evenly spaced, like bulbs strung regularly along a highway.

But when you asked all the positive rationals to light up, they turned out to be so densely packed that despite the fact that all the irrationals remained off, you could no longer see any gaps. The line now looked like a gleaming strand of unbroken illumination.

Surely this means the rationals have a higher cardinality? Or are you missing something?

33.

BATTLE OF THE CARDINALITIES

RATHER THAN HAVE HIM RUMMAGE AROUND IN THE Vatican library, I'm going to save the pope a trip. The person he's thinking about is Georg Cantor, who did, indeed, send his mathematical work to Pope Leo XIII. Since the pope wondered about Cantor's correspondence, let me briefly fill him in.

Georg Cantor was the first who thought about assigning a cardinality to the naturals. The idea was disparaged by several fellow mathematicians because of the very issue we encountered earlier: can infinity really exist, except in its potential form? The battles Cantor fought, grappling with both infinity and the criticism of his colleagues, took their toll. He suffered bouts of severe depression, needing a series of hospitalizations for his breakdowns. What kept him going was religious faith. He felt God was personally revealing the secrets of infinity to him, which gave him the conviction that his work eventually had to win the acceptance it deserved.

Since infinity is so often tied in with interpretations of divinity and God, Cantor was eager to have his theories blessed by the Catholic Church. In particular, he didn't want to become a mathematician version of Galileo, accused of blasphemy for competing with the theological version of infinity. He corresponded with several religious scholars and clergymen, and finally got a cardinal, no less, to certify that his work was not at odds with Christian theology. Pope Leo XIII's efforts to reconcile science with religion had set the stage for this unusual interaction; Cantor wrote to the pope himself in 1897.

You'll notice I've named the mathematician in my interplanetary tale Georg, after Cantor. The planets' names also come from him. Aleph (ℵ), the first letter of the Hebrew alphabet, is what Cantor used as the symbol for the cardinality of the naturals—he denoted it $\aleph_0$ (for which we've been using ∞). Cee just stands for the letter *c*. We'll shortly see what role it plays.

Let's go back to Georg's tale. The pope adores parables, which he senses this is turning out to be.

Georg saves the day again

The Ceenums rise once more. This time they gird for battle with a phalanx of lasers that appear as numerous as the rationals. Looking through the telescopes, the Alephites observe that each laser is now marked with a positive fraction. (The Ceenums figure that's going to be enough, so they haven't bothered with the negative fractions or zero.)

By the time Georg gets the news, the planet of Aleph is in full hysteria mode. The teeming assemblage of Cee's lasers look so fearsome, some say, that scouts at the telescopes have fallen down, stricken. Aleph's outdated arsenal, frozen in time since the naturals (the factory has still not been repaired), can hardly hope to offer much resistance when confronted with the rationals. Surely this spells the end of the planet's freedom.

Georg notices that one of Cee's lasers is marked 1/2, another 2/4, and yet another 3/6, even though these fractions all reduce to the same rational. Watching the Ceenums prepare for attack, he realizes why: this allows them to form a neat array. Lasers whose fractions have numerator 1 take the first row, numerator 2 the second, and so on. A similar denominator-wise arrangement appears in the columns.

1/1	1/2	1/3	1/4	1/5 ...
2/1	2/2	2/3	2/4	2/5 ...
3/1	3/2	3/3	3/4	3/5 ...
4/1	4/2	4/3	4/4	4/5 ...
5/1	5/2	5/3	5/4	5/5 ...
⋮	⋮	⋮	⋮	⋮

Georg wonders how he might use his meager one-dimensional list of naturals to neutralize this vast two-dimensional array. If he starts at the top left and performs his matching from left to right, he'll face an obvious problem: he'll never make it to the next row, since all the naturals in {1, 2, 3, . . . } will be used up.

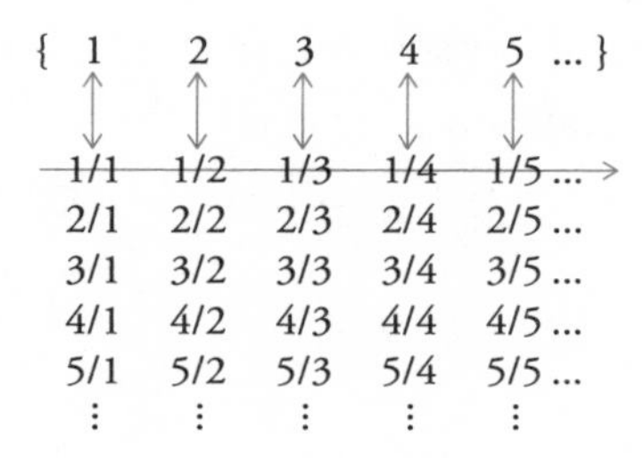

The same problem will occur if he tries matching the array by *columns*—starting with the first, and working downward, he'll never make it to the second.

1 ⟷	1/1	1/2	1/3	1/4	1/5 ...
2 ⟷	2/1	2/2	2/3	2/4	2/5 ...
3 ⟷	3/1	3/2	3/3	3/4	3/5 ...
4 ⟷	4/1	4/2	4/3	4/4	4/5 ...
5 ⟷	5/1	5/2	5/3	5/4	5/5 ...
...	⋮	⋮	⋮	⋮	⋮

What magic numbering scheme will allow him to squeeze two dimensions into one? Georg tries not to think of the possibility that this might be just plain unachievable. If the cardinality of the rationals is indeed higher than ∞, then Aleph is surely lost.

And then, just when the planet elders are about to signal their acquiescence, Georg has it! Rather than address the Ceenums' laser array by row or column, he should do so along successive *diagonals.*

In other words, starting with 1/1, he should next proceed along the diagonal comprising 2/1 and 1/2, then along the next diagonal of 3/1, 2/2, 1/3, and so on.

1/1	1/2	1/3	1/4	1/5 ...
2/1	2/2	2/3	2/4	2/5 ...
3/1	3/2	3/3	3/4	3/5 ...
4/1	4/2	4/3	4/4	4/5 ...
5/1	5/2	5/3	5/4	5/5 ...
⋮	⋮	⋮	⋮	⋮

Each diagonal is finite, so starting from the top left, he can march through the entire array, covering it all. This will allow him to engage every item in Cee's array with a unique natural, thereby neutralizing the entire formation!

Ceenums {	1/1	2/1	1/2	3/1	2/2	1/3	4/1	3/2 ... }
	↕	↕	↕	↕	↕	↕	↕	↕
Alephites {	1	2	3	4	5	6	7	8 ... }

And that's how Georg saves his planet again—by putting this matching strategy into action. The Alephites heave a sigh of relief, and thank him—he's their new patron saint.

Let's take a closer look at Georg's trick. If we write (1/2) as (1,2), (2/3) as (2,3) and so on, then what we've really done is establish that the set of all ordered pairs of natural numbers has cardinality ∞. Now if you delete repetitions like (2,4) and (3,6), which both correspond to the fraction (1/2), that is, the pair (1,2), then the cardinality can't increase—in fact, it remains ∞, or, in Cantor's notation, $\aleph_0$. Since the negatives can be treated similarly,* the set of rationals is equal in

* Recall how the negative integers were alternated with the positive ones so that the entire set of integers could be matched with the naturals. A similar alternating strategy can be followed again for the negative and positive rationals. Zero can also be tucked in.

cardinality to the set of naturals. So there are just as many fractions as there are whole numbers! This might seem incredible, but we've demonstrated it's true.

Which raises another question. What about the reals? Can we show that their cardinality is *also* $\aleph_0$? In other words, that there are exactly as many reals as there are naturals? If so, this would answer our question about ordering all the individual time frames between any two instants—we could just number them the first, the second, the third, and so on. The continuum would be just another list, its mystery would be solved!

34.

A SMILE ON GEORG'S FACE

AFTER ALL THESE RESCUES OF THEIR PLANET, THE Alephites become blasé about their defense. Their arsenal of naturals has stood the test of time, countered every army the Ceenums have been able to create. There has been no attempted invasion for years now—surely their adversaries have given up the idea of attacking again. In any case, with Georg on their side, they'll be able to figure out the correct arrangement to address any future threat. The project to repair their sabotaged laser facility falters, then comes to a standstill. There are so many other deserving initiatives—the money is soon budgeted elsewhere.

Georg watches these developments with dismay. He tries to warn his fellow Alephites that they've lulled themselves into a false sense of security. The reason the Ceenums haven't returned is cause for great alarm, because it means they are planning a much broader offensive. In fact, he is almost certain about what this will entail—an army of lasers that equals the reals in cardinality. The reason it is taking so long is the sheer scale of manufacturing involved.

But everywhere he goes, the Alephites dismiss his fears. They tell him he's being paranoid, that he should not be afraid to accept the blessing of peace bestowed on them. "Why not enjoy it, instead of being so hawkish?" They laugh off the seriousness of the threat posed by the reals. "Whatever the danger, you'll manage to find the right arrangement of the naturals to counter it. We have full faith in your abilities. You should, as well!"

Although Georg is getting increasingly jittery, the confidence his compatriots express in him offers a ray of hope. Perhaps they're correct, perhaps he's too pessimistic. He decides to further explore the issue by simulating a Ceenum attack, playing a war game of sorts. Suppose there were enemy lasers pointed at them, with cardinality equal to that of the reals. How would he counter them with just the natural number arsenal at his disposal?

He tries refocusing the defense lasers in clever ways, gathering together blocks of neighboring reals, using their decimal expansions. He dreams up tricky matching algorithms based on the naturals' defining sets. But nothing works. The key for success is for the reals to have the same cardinality $\aleph_0$ as the naturals. Is this true? He can't convince himself.

After many failed attempts, he tries a different tack. He'll *assume* this proposition is true and see whether it leads to any contradictions.* If it does, the statement must be false, and the reals will overwhelm the naturals in any match. But he hopes no inconsistency will emerge, bolstering his confidence in being able to repel a Ceenum attack.

So he imagines the reals can be put in one-to-one correspondence with the naturals—in other words, can be counted. As a first step, he starts with something easier: he assumes only that those reals between 0 and 1 (which constitute a smaller set) can be thus listed. Such a list would give him a first real, a second real, and so on—an ordering that could be followed in aiming his defense lasers to get parity and head off an attack.

Now each such real number x has a decimal expansion that begins with 0 (for instance, $x = 0.4358133907\ldots$). Georg imagines that each such expansion has been matched with a corresponding natural. To get a better feel for this, he randomly selects some reals that might make up such a list, writing down their first several decimal digits.

---

* This is a common approach in mathematics, called *proof by contradiction.*

$$
\begin{aligned}
x_1 &= 0.4\,3\,5\,8\,1\,3\,3\,9\,0\,7\ldots \\
x_2 &= 0.7\,1\,8\,9\,1\,7\,5\,1\,2\,8\ldots \\
x_3 &= 0.6\,9\,8\,1\,9\,0\,0\,1\,5\,7\ldots \\
x_4 &= 0.3\,2\,8\,5\,6\,2\,2\,8\,9\,5\ldots \\
x_5 &= 0.5\,0\,0\,0\,0\,0\,0\,0\,0\,0\ldots \\
x_6 &= 0.4\,5\,6\,2\,4\,5\,6\,2\,4\,5\ldots \\
x_7 &= 0.9\,9\,9\,7\,7\,3\,6\,8\,8\,2\ldots \\
x_8 &= 0.1\,6\,3\,5\,8\,2\,1\,1\,6\,5\ldots \\
x_9 &= 0.0\,4\,1\,4\,1\,4\,1\,4\,1\,4\ldots \\
x_{10} &= 0.2\,7\,9\,2\,8\,1\,1\,9\,5\,2\ldots \\
&\ldots\ldots\ldots\ldots\ldots\ldots
\end{aligned}
$$

Georg stares at his list a long time. Gradually, he starts sensing something amiss. An inconsistency flutters at the edge of his consciousness—he glimpses a whirring of wings. And then, he's able to capture it. There's a way to create a real number y between 0 and 1 that doesn't match any of the numbers in the list!

Excited by this prospect, even though it might spell doom for his planet, he starts writing down y's decimal expansion. Since the first digit of x_1 after the decimal is a 4, he'll take the first digit of y to be something different: a 1.

$$
\begin{aligned}
x_1 &= 0.\textcircled{4}\,3\,5\,8\,1\,3\,3\,9\,0\,7\ldots \\
y &= 0.\,1\ldots\ldots\ldots\ldots\ldots\ldots
\end{aligned}
$$

Similarly, for the second digit of y, he'll look at the second digit of x_2 and take something different. Since this second digit of x_2 is a 1, he sets the second digit of y to be a 2.

$$
\begin{aligned}
x_1 &= 0.\textcircled{4}\,3\,5\,8\,1\,3\,3\,9\,0\,7\ldots \\
x_2 &= 0.7\,\textcircled{1}\,8\,9\,1\,7\,5\,1\,2\,8\ldots \\
y &= 0.\,1\,2\ldots\ldots\ldots\ldots\ldots
\end{aligned}
$$

After this, he simply repeats this strategy. For subsequent digits of y, he looks at his list: at the third digit of x_3, the fourth digit of x_4, and so on, changing each digit to 1 or 2 as he goes along.

$$
\begin{aligned}
x_1 &= 0.\textcircled{4}\,3\,5\,8\,1\,3\,3\,9\,0\,7\ldots\\
x_2 &= 0.\,7\,\textcircled{1}\,8\,9\,1\,7\,5\,1\,2\,8\ldots\\
x_3 &= 0.\,6\,9\,\textcircled{8}\,1\,9\,0\,0\,1\,5\,7\ldots\\
x_4 &= 0.\,3\,2\,8\,\textcircled{5}\,6\,2\,2\,8\,9\,5\ldots\\
x_5 &= 0.\,5\,0\,0\,0\,\textcircled{0}\,0\,0\,0\,0\,0\ldots\\
x_6 &= 0.\,4\,5\,6\,2\,4\,\textcircled{5}\,6\,2\,4\,5\ldots\\
x_7 &= 0.\,9\,9\,9\,7\,7\,3\,\textcircled{6}\,8\,8\,2\ldots\\
x_8 &= 0.\,1\,6\,3\,5\,8\,2\,1\,\textcircled{1}\,6\,5\ldots\\
x_9 &= 0.\,0\,4\,1\,4\,1\,4\,1\,4\,\textcircled{1}\,4\ldots\\
x_{10} &= 0.\,2\,7\,9\,2\,8\,1\,1\,9\,5\,\textcircled{2}\ldots\\
y &= 0.\,1\,2\,1\,1\,1\,1\,1\,2\,2\,1\ldots
\end{aligned}
$$

Using his recipe, Georg is quickly able to write down several digits of a y corresponding to his sample list. Were he to take more numbers in his list, he'd get more digits of y—a million, a billion, whatever—a long sequence of 1s and 2s. If he starts with an infinite list of x's, this decimal expansion will be infinite, so that y will be completely determined.

Here's the crucial point. The way he constructs y, it can't equal any number on the list! Its first digit (which is 1) doesn't match the first digit of x_1 (which is 4), so clearly $y \neq x_1$. Also, y can't equal x_2 because their *second* digits don't match (y's is 2, x_2's is 1). In the same way, y can't be x_3 or x_4 or, in fact, *any* of the other x's, because due to our construction, there is always at least one digit that doesn't match. So y is a number between 0 and 1 that isn't on the list, even though every number between 0 and 1 *has* to lie on the list.

This is a contradiction. The source can be only one thing—the assumption Georg made of the reals having the same cardinality as the naturals. This assumption has to be *false*. There will always be at least one real number left over, no matter how you try to match them with the naturals.

For a brief moment, Georg feels pure, unadulterated euphoria over this realization. He's figured out the conundrum! There are more reals than naturals!

But then, a deep, shuddering fear clamps down over his heart. They will be defenseless against a Ceenum army as numerous as the reals. No matter how he aims his arsenal, there will always be at least one enemy laser left over, one that can't be neutralized. A laser with the capability to blow up Aleph.

Galvanized by this knowledge, Georg rushes out to warn his fellow citizens. He visits labs and universities, approaches every higher-up he can, even accosts pedestrians on the street. But few take the time to understand his proof, and the handful that do, dismiss it. The fact that he has such a desperate look in his eyes, appears so haggard and feverish, doesn't help. His attempts to get the abandoned laser plant up and running again earns him the reputation of being "another one of those crazy mathematicians."

Eventually, he *does* go over the edge. He suffers a severe nervous breakdown and is hospitalized. The mayor himself comes to visit. All Georg can do is babble about cardinality and laser production. The mayor pats the hand of this former hero for the benefit of the press, but the photo opportunity he hoped for is a disaster. As he leaves, he gives instructions for Georg to be put in an institution.

Things get worse. Unable to convince anyone of his message, Georg takes out his frustrations on the staff. He is confined to increasingly restrictive environments, and eventually finds himself in a straitjacket, tied up for his own good. He tries to withstand the erasing effects of forcibly administered drugs by hanging on to the one thought that has become key to his existence: the cardinality of the reals. "Greater than the naturals, you'll see," he takes to spouting, to nobody in particular. In time, this becomes a mumble.

And then, one year, the Ceenums launch their offensive. As Georg predicted, they assemble an army as numerous as the reals. In the ensuing chaos, the nurses and orderlies flee, and Georg is able to wander out of his room to the grounds outside for the first time in a long while. The institution is built on a hill, and he looks at the city below, where sirens blare and the streets are panicked. Looming over them all is the planet Cee, a giant inflamed pupil in the sky, closer and more malevolent than he can remember ever seeing it.

Georg nods to himself. He can no longer articulate his mantra, though the word "cardinality" pops up in his mind. He nods again, and the muscles in his face twitch. He tries to remember how to pull his lips into a smile, but it comes out a grimace.

Thus concludes the parable of Georg and the Alephites.

The cardinality c

The pope's unsettled by the ending, as are you. Maybe it turned out more harrowing than I intended. That's probably due to the way things are going these days, with so many ignoring what mathematicians and scientists advise. About everything from pollution to virus vaccinations to climate change—you know the list. At our collective peril, needless to say. In case you're wondering whether the Alephites survived, they didn't.

My hope is you've been scared straight. Into restoring logic and reasoning back to their rightful place. Paying attention to what mathematicians say. That's the reason, after all, I'm writing this. Perhaps the pope will agree to be my fellow evangelist. We'll tour the planet together to bring about change. To save the Earth from Aleph's fate.

Before embarking on this noble pursuit, though, let's examine the other revelation that emerged from my tale. What exhilarated Georg so much, even as he looked into the void. It's the fact that ∞ isn't the only infinity there is!

Here's what establishes this. Matching the reals between 0 and 1 to the naturals, as Georg tried, ends up in at least one real left over. So the cardinality of the two sets can't be the same; that for the reals has to be *greater* than that for the naturals. Call this cardinality c (the c of planet Cee), then

$$c > \aleph_0$$

This means the naturals, integers, and rationals all have the same cardinality $\aleph_0$, but the reals are more numerous.* The points in a line segment and the instants in a time interval both correspond to this higher infinity. Which means, in answer to the question we posed earlier, that these *cannot* be ordered as the first, the second, the third, and so on. The infinity we move through between points, or live through between instants, cannot be counted in the same way that naturals can. We get to experience not just countable infinity $\aleph_0$, but a higher, uncountable infinity c in our lives!

This brings us to another question. Could there be an infinity even greater than c? A set of numbers even more numerous than the reals?

* Since the reals simply combine the rationals and the irrationals, and we already know that the rationals have cardinality $\aleph_0$, the irrationals must have cardinality c. This means the irrationals are infinitely more numerous than the rationals, and the real line is mostly filled with irrational numbers! A surprising revelation, indeed, given how rarely we encounter irrationals in our lives.

35.

THE END OF MATHEMATICS

IN THE HUNT FOR A HIGHER INFINITY, LET'S FIRST note that the infinity we called *c* was the cardinality of only the points between 0 and 1. That's what Georg showed. Could the cardinality of the entire real line, stretching from minus infinity to plus infinity, be higher? After all, this line can be thought of as a unit interval (like the interval from 0 to 1) repeated over and over again.

The answer is no. The set of all real numbers also has cardinality *c*. There's a neat geometric way to see this—we can put the points between 0 and 1 in a one-to-one correspondence with the positive real axis. Simply bend the unit interval into a quarter circle with center A. Then match each white point on the arc (there are *c* of them) with a black point on the line as shown. As the white points rise up toward the end of the arc, the corresponding black points advance further along the line. You can see that the numbers of white and black points have to be exactly the same—it's like matching the fingers of two hands again. This means the finite interval 0 to 1 has exactly as many points as an infinite line! In other words, their cardinalities both equal *c*.

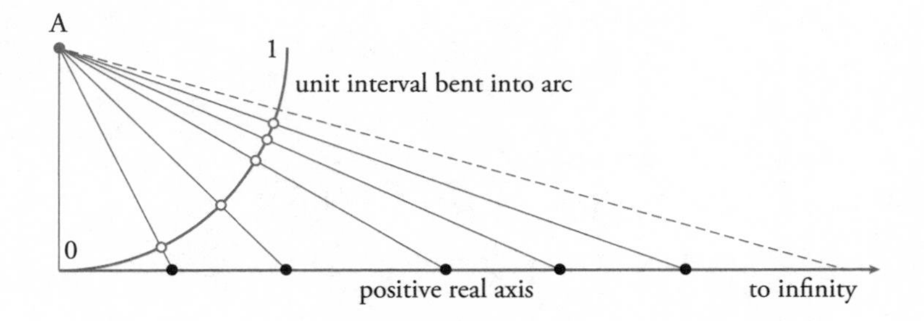

To continue with the hunt for a higher infinity, what about the complexes? Alternatively, all the points in the plane? As we've seen, we can represent both these sets by ordered pairs (x,y) of reals. Surely these should be more numerous than the reals on a single axis?

They're not. We saw just recently that taking pairs doesn't give you a higher cardinality, at least for the case of x and y natural. Recall how Georg managed to put ordered pairs (x,y) of naturals in one-to-one correspondence with just a single list of the naturals (by advancing along diagonals). The same kind of equivalence, it turns out, can be shown for (x,y) real—but only via a different, more involved argument. In fact, ordered triples of reals don't raise the cardinality either, and neither do ordered quadruples. So both 3-D space and 4-D spacetime will have the same cardinality c. This c truly is the cardinality of any continuum.*

That pretty much exhausts all the sets we've created so far. Are we ready to throw in the towel, then? Recognize c as the biggest infinity there is?

If so, there would be a certain closure in this. We could feel reassured that we'd at least reached the outer boundary of mathematics, even if it enclosed much uncharted territory left to explore. Also,

* Incidentally, this answers a question posed earlier, when i and the rest of the complexes were being uprooted from the plane. Could they all have been squeezed into just the condos on the real line? The answer is yes, since they have the same cardinality as the set of these condos. But there wouldn't be any way to do so that preserved the orderliness of the real line, so although guilt inducing, their eviction was still necessary.

we could feel confident that the universe we designed was the right one, given how all our math was being showcased in it, with nothing wasted. Even the different infinities would each have been pressed into service: the $\aleph_0$ kind in counting, and the *c* version in continuums of time and space. With this milestone question of infinity settled, we could even perhaps start to wrap up our thought experiment.

Alas, it's a milestone we haven't reached yet. Rather than being the maximum, *c* turns out to be just one in a whole hierarchy of infinities.

The game of subsets

To access an infinity greater than *c*, consider the following question: Given a set of numbers, how many subgroups—or "subsets"—can you construct from it? For instance, if you start with the empty set Ø (whose cardinality is zero), you can get exactly one subset: Ø. Start with the set {1}, and you get two subsets: Ø and {1} itself. As the cardinality of your original set increases, so does the cardinality of all the subsets you can form from it. For instance, with {1, 2} you can form the subsets Ø, {1}, {2}, and {1, 2}, whereas with {1, 2, 3} you can form the subsets Ø, {1}, {2}, {3}, {1, 2}, {1, 3}, {2, 3}, and {1, 2, 3}.

Look at this systematically, and you'll realize that the number of subsets you can form is always greater than the cardinality of the set itself. *One* subset formed when you start with a set of cardinality zero, *two* subsets formed when you start with a set of cardinality one, *four* subsets starting with cardinality two, *eight* subsets with three, and so on. *Always* a larger number.*

Here's the punchline. You can prove that the number of subsets is greater no matter what the cardinality of the set you start with is. Even when this starting cardinality is infinite! (I've included Cantor's ingenious proof for this in the endnotes.)

* In fact, starting with a set of finite cardinality n, the number of subsets you get is 2^n, which is always larger than n.

Now, this is not a problem if you just start with the infinite set of naturals (or rationals). Such a set has cardinality $\aleph_0$, and you can show that the cardinality of all its subsets is greater—in fact, it turns out to be exactly *c*. But what if you consider all possible subsets of the reals? You'll now get a set whose cardinality has to be greater than *c*. In other words, you'll get a higher infinity! If we use the labels $\infty_1 = \aleph_0$ (the cardinality of the naturals) and $\infty_2 = c$ (the cardinality of the reals), then we get a new infinity, which we label ∞_3 (the cardinality of all subsets of the reals), which is larger than ∞_2!

This infinity will exceed the number of points on a line or plane, the number of points in 3-D space, and even the number of points in 4-D spacetime. It's something new in our thought experiment, neither experienced nor seen before. It's hard to think of how we might be able to get Nature to ever use it.

But that's not all. Starting with this set of cardinality ∞_3 (i.e., the set of all subsets of the reals), and then finding all subsets of *that* set, would give something with an even *higher* cardinality. And repeating this process of taking subsets over and over again would yield an endless hierarchy of infinities:

$$\infty_1 < \infty_2 < \infty_3 < \infty_4 < \infty_5 < \ldots$$

Trying to put a face on all these infinities through our physical universe would surely be an impossible dream. Though for all we know, perhaps there's some completely different version of the universe we could have designed to perfectly embody them. In the same way that physical infinity was part and parcel of the planets Aleph and Cee.

In any case, the picture we have to contend with is a range of escalating summits looming over our mathematical landscape, endless and awe inspiring. The universe of mathematical ideas, which we were so close to congratulating ourselves on having fully conquered, is suddenly rife with unexplored possibilities. There can never be a final frontier in this universe; each goalpost simply leads to another, visible in the distance.

The continuum hypothesis

It isn't just the boundary of mathematics that turns out to be illusory, there are also hidden valleys closer in that we cannot ever scope. Infinity harbors the following riddle in its folds. Recall that $\infty_1 = \aleph_0$ and $\infty_2 = c$. Could there be an infinity in between these? Could we build a subset of the real numbers that is neither as sparse as the naturals, nor as dense as the continuum, but has an intermediate cardinality?

Cantor battled this question a long time. Several times, he thought he had it figured out, that the answer was no, but then a mistake in his reasoning would pop up. At other times, he'd be sure he had come up with an example of precisely such a set, only to find he'd erred once more. The inability to answer this question caused acute depression, even hospitalization. In the end, it led to his giving up mathematics completely.

Little did Cantor know that his efforts were doomed from the start, since he was pursuing a statement that could not be proved either true or false! The *continuum hypothesis* is the assumption that there is no intermediate infinity, that is, c is the next possible cardinality after $\aleph_0$, with nothing else in between. Assuming this statement is true is perfectly compatible with the rest of mathematics, but so is assuming it's false; neither assumption leads to a contradiction. In other words, the proposition is *undecidable*—we cannot use the tools and logic of our underlying mathematical system to establish or refute its trueness. In 1931, the Austrian mathematician Kurt Gödel proved there will always be such undecidable statements in many systems of mathematics. In other words, these systems can never resolve every question posable in them*—they will be *incomplete*.

These two final characteristics of infinity—its endless hierarchy

* We saw a good example of such undecidability on Day 2: the parallel axiom cannot be either proved or disproved based on the remaining assumptions for geometry set up by Euclid.

and continuum undecidability—do not impinge on the work of most mathematicians. Nor are they results we'll need to teach Nature, as part of her practical training. And yet, they hold valuable insights for us, much like the other messages we've extracted from numbers—such as the concept of randomness or the motivation for geometry.

The first insight pertains to perfection. Many might agree with the ancient Greek philosophers who believed that mathematics is the most perfect system there is. Everything's based on well-defined rules, results are logically derived, there are no material or corporeal considerations that might introduce flaws or variability. And yet, the system doesn't give us complete knowledge, is what Gödel seems to be saying. There will always be questions that can be posed whose answers we cannot prove. Even with such perfection, we cannot hope to get to the root of everything.

Let me be clear that Gödel's work only addresses the incompleteness of certain well-defined mathematical systems; trying to apply his theorem to other fields, or to "prove" that life in general can't be flawless, is nonsense. However, maybe one shouldn't hope to find perfect knowledge anywhere if this fails for even the most promising candidate: mathematics. So don't set expectations too high for other systems of thought (like religion or physics) either, don't be too surprised or disappointed if they fail to explain everything. As for omniscience, it's a flat impossibility within Gödel's incomplete mathematical systems, since you won't know the proof of everything. The pope will be relieved to know, however, that Gödel says nothing about the omniscience of an external Supreme Being.

The multitudes of infinity also speak to us. There's something reassuring about the vast reach of mathematics, about its unending series of questions to keep us engaged, exploring. We can't hope to cover this entire landscape or even know in advance which tracts might prove useful—not everything will, as all those endless infinities attest. So far, the model in our narrative has been to create mathematics and then apply it right away. This is fine for the basics, but now we risk getting lost in the endless terrain ahead. We'll need to be more focused and develop new math to address specific issues encountered. Otherwise, we may not find a use for it.

Might we leave it to Nature to create such mathematics herself as she builds the universe? We have reason to be optimistic, given the ingenuity she displayed in using randomness to generate a sphere. Our past fears of her getting too independent or going rogue were, I'll admit, misplaced. We'll take her as far as we can, of course, provide her with stacks more blueprints. But we can't possibly anticipate every issue that might arise once construction begins—the devil is going to be in the details. Nature will have to come up with quick solutions herself at the front lines if the universe is ever to be completed.

There will be many objections, I'm sure, to handing her the reins. We're talking of doing so before anything tangible exists, before fingers or toes are manifested. Will she be up to the task? Has she been properly vetted?

These are not the right questions to ask. The real issue is whether we've taught Nature enough, so she can develop mathematics as needed. To which the answer is yes—we've begun with zero, and gone all the way to infinity. We've started Nature off on the founding properties of numbers, geometry, algebra, physics. That's the knowledge that enabled us to come this far, so why wouldn't it also be enough for her to proceed?

There comes a point while building any house when you nervously give your contractor control. You go through the plans one last time, check out that she has the right assistants and tools. After that, it's in her hands; there's nothing left to do but wait. You hope Nature's up to the task, that she'll be able to deliver the universe you want.

Day 7

EMERGENCE

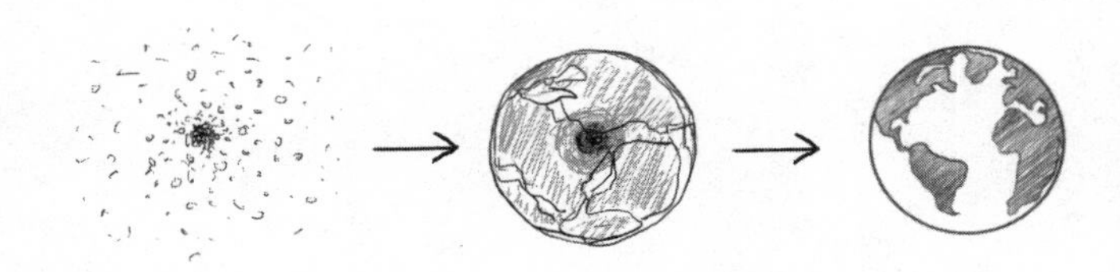

Launching the iterations
from which everything will emerge

36.

SETTING NATURE FREE

NATURE'S MODUS OPERANDI IS MUCH MORE HANDS-on than yours. Rather than continue with more planning and simulation, she intends to go ahead with a physical launch. She's keen to start with something tangible, molding it as she goes along.

Word of the launch gets out. A large audience gathers to watch. Even the pope tunes in, via Zoom, from Rome. With his beaming smile and pressed white cassock, he looks positively radiant. And why not? How many pontiffs have witnessed the start of the universe?

Several of the spectators are physicists. You can tell by the dark glasses they wear, as if they're expecting an eclipse, and by the scientific instruments they carry. One has even lugged along a Geiger counter. They're abuzz with excitement, emitting a continuous low murmuring that does, indeed, resemble the sound of bees.

Nobody's quite sure what to expect. Will there be a countdown, like before a rocket launch? Will space emerge, gravity form? Will the skies fill with magnificent comets, the rounded surfaces of water-covered planets? Will Nature alert everyone to the first flicker of life being born?

But Nature's nowhere to be found. She's busy battening down the hatches in preparation for what's to come. Figures, symbols, equations, infinities big and small—she's gathered them all up to take along. The numbers are packed so tightly within her folds that she could be Noah's ark.

She suddenly notices the spectators. Does she owe them a show if they've come this far to watch? Perhaps she could serve each group only what they want. Special effects for the believers, primordial soup for the big bangers. With a side dish of fractals for all.

The thought of satisfying an audience makes her blanch. Is she ready? Can she make it come out right? There's so little nailed down, despite the reams of blueprints you've supplied. Has she even grasped the full enormity of what's being asked?

But then she feels the reassurance of the numbers. They're one with her, they'll be embedded in the fabric of her universe from the start. She's assimilated enough mathematics by now to face any challenge to come. The universe will be an opportunity to show all she's learned.

Exhilarated by this thought, Nature takes the plunge.

If at first . . .

I'd love to end here, with this metaphorical image of Nature, armed with mathematics, poised at the threshold of creation. But it leaves too much dangling. Readers may accuse me of copping out if I don't connect the dots to where we are. How well would Nature succeed in creating our particular universe, given her point of launch?

To be frank, she almost certainly *wouldn't* succeed. There are too many unknowns, too much left to design and coordinate. Even if she successfully activated gravitation, say, each additional phenomenon that emerged—like heat or electricity or magnetism—would need more formulas. Picking these correctly would be a challenge due to the way everything is connected in our universe. For instance, the earth's *magnetic* field arises from *electric* currents generated as a result of *heat* escaping from the core, which is held together by *gravity*. Getting even a single relationship wrong would throw off the whole system. Recall how many different ways there are just to incorporate time into our model.

Suppose, though, that this was all taken care of by Hilbert's helpers, that the blueprints they provided were exhaustive enough to characterize every last connection. That still wouldn't do it. Multitudes of physical constants (e.g., the intensity of gravity, the mass of an electron, the speed of sound) would have to be fine-tuned in the formulas. Even small deviations could cause the universe to break down. For instance, if gravity was stronger, everything would scrunch together into one big ball; if weaker, planets might not form. We have no magic hat from which to pull out the precise value of every constant in advance.

So, trying to predesign the universe to get it exactly as we know it is like throwing a dart at a bull's-eye on Mars. There's just about zero chance it will work.

Suppose we ask a broader question: Could we expect Nature's launch to lead to *any* coherent universe at all? The answer remains no. There's still too much to coordinate and fine-tune—the bull's-eye might have moved closer, but only to the moon instead of Mars. Again, Nature's attempt would almost surely fizzle out.

One way to better the odds would be to allow Nature multiple tries. Then she could make small changes after each failure—either randomly or based on some feedback algorithm (for instance, one that uses past data on planets jelling or not to modulate how much to crank gravity up or down).* We could also let her make such adjustments at junctures along the way, without having to relaunch. Her series of successive attempts would constitute something familiar: an input/output process.

Recall these were the mechanisms behind coastlines forming—the output from one cycle becomes the input for the next, each shaped by similar forces of erosion. Taking the place of such forces now would be Nature's changes; instead of a coastline we'd have the universe's successive drafts. Most of these iterations would still go nowhere. But every once in a blue moon, perhaps the right confluence of conditions would lead to a viable universe.

---

* As we'll show ahead, such feedback or learning loops can proceed autonomously, without Nature consciously manning the controls.

Be prepared for such a universe to be dramatically different from ours! As Nature proceeds with her incremental changes iteration by iteration, she might be led to some entirely new combination of laws that work. She could end up defining distance differently—favoring the zigzag routes of taxicabs over the direct flight paths of crows. Or framing everything in an alternative geometry, choosing from a smorgasbord of selections (flat vs. curved, finite vs. infinite, 3-D vs. 4-D or higher, shaped like a sphere or doughnut or even pretzel, why not?). She might find new ways to distribute matter in space—for instance, in a uniform grid, or as a giant 3-D cosmic fractal.* Instead of setting a ceiling on maximum speed, she might impose such a restriction on some other quantity (volume? mass?), with peculiar consequences. As for the free-willed simulated particles, who knows what properties she'd end up bestowing on them to buy their compliance?†

Is there any evidence for any of the above? The repeated tries, the iterative chains, the successes that look nothing like ours, the plethora of rejects along the way? Are there multiple universes actually hanging around somewhere?

Theoretical physicists might say yes, pointing to the so-called multiverse. Such a construct, consisting of numerous parallel universes, can allow *all* chains of possibilities to be realized, along multiply branching paths. For instance, a particle faced with twenty slits could simultaneously pass through each of them, but in twenty separate

---

* Such fractal distribution isn't a fanciful idea, but a matter of serious debate pertaining to our own universe. Is matter distributed randomly in space or does it display self-similarity at different scales in some statistical sense?

† Maybe that's how the elementary particles in our own universe secured some of their more outlandish properties at the quantum level. For instance, their ability to pass through two apertures simultaneously when an obstacle is placed in their path, their power to transmogrify into an alternative wave-like avatar at whim, their ability to instantly communicate with "entangled" twins hundreds of miles away. Could these capabilities—all of which have been experimentally observed, mind you—be bribes in return for giving up free will? (Perhaps there's a great science fiction story waiting to be written with this premise.)

universes. Similarly, physical constants could assume a range of values, with a separate universe for each combination.

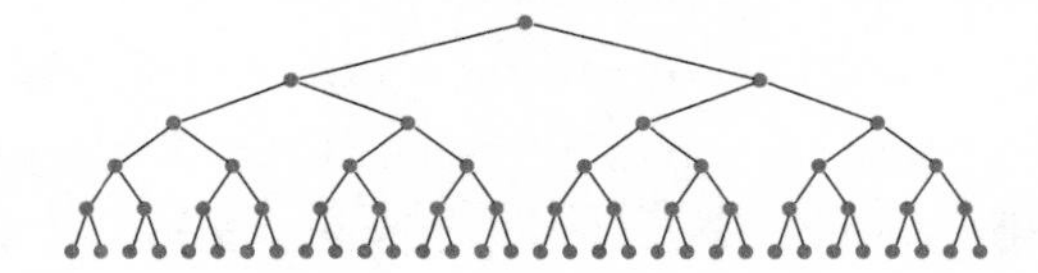

The obvious problem is that the number of universes needed to accommodate all such strings of choice quickly becomes untenable. Also, such theories, being unverifiable, do not rise to the level of traditional experiment-based science. The multiverse is controversial among physicists, one of whom once dismissed it to me as not science but science fiction.*

Fortunately, we're not physicists, so we don't have to pick a side in this argument. Considered as mathematics, input/output rules and any resulting universes are perfectly acceptable concepts. In our thought experiment, Nature's free to iterate to her heart's content.

When she does, her efforts can bear fruit in two ways. The first is that Nature succeeds so often (iterating from multiple launches) that she creates just about every universe possible, including ours. This would explain how our universe came about. But to take care of all these possibilities, we'd need to assume Nature does an awful lot of construction. We're talking $\aleph_0$ or c or an even higher infinity's worth of universes, so perhaps this isn't our thought experiment's most satisfying conclusion.

---

* There is a sense among some physicists that what some of their colleagues are doing is not physics but mathematics. More specifically, that several theories in particle physics (and related fields) are being proposed because they follow from the mathematics, rather than being based on experimental evidence. We encountered this effect earlier, when superstring theory required space to be nine-dimensional. Keep in mind, though, that math and physics are inextricably linked—so it's not always possible to separate them. Also, when an axiomatic system is used to build up an area of physics, such theoretical inferences will naturally be encountered. As a mathematician, I'm certainly open to exploring theories that capture the various possibilities reality can follow.

The second possibility is that Nature has limited stamina (and luck)—so she succeeds in rendering only a limited number of universes. Let's be extra stringent and put this number at one, reflecting the fact that the only universe we've actually observed is our own. Then we're faced with the obvious puzzle: Out of the vast multitude of universes possible, why would Nature have stumbled onto creating ours?

The answer: it's really a perceptional illusion. Our universe might seem exceptional to us, but that's only because we happen to be in it, we've been acculturated to think of it as such. To a neutral observer, it's probably rather ordinary—quite interchangeable with a bunch of others. No matter *which* universe Nature ended up creating instead, she'd face the same issue. A life-form on it would think it just as special, believe it to be the result of extraordinary coincidence. Even though, from a cosmic viewpoint, nothing out of the ordinary had happened.

Think of it in terms of the lottery. There's going to be a winning combination chosen no matter what, with one set of numbers as likely as another. Look at today's winner, and you'd just shrug it off as a random, totally unremarkable combination of numbers. But if it's the combination you bought, then you'd think a miracle had occurred.

So, yes—iteration might be just the thing to give Nature a plausible route to our universe.

Life and mathematics

The pope has been rather silent, even though I'm certain he doesn't agree with all I've said. If Nature is supposed to be God at work, then quite a few of my assumptions are problematic. I'm expecting a summons from the Vatican any day now, so had better prepare my defense. To which of my points is the pope likely to object?

For starters, the whole idea of Nature iterating. God doesn't need to iterate, the pope will say—He can simply aim for perfection, achieve it in one shot. The idea that a Supreme Being has to work His way through a series of drafts is absurd.

In response, I'll point out that iteration doesn't contradict religion; there's no reason it can't be employed in the divine scheme of things. Take the successive days of Genesis: Surely they indicate an iterative frame of mind? Other religions, like Hinduism, are more explicit—for instance, Brahma blows out the universe again and again. As for drafts, with so many universes imaginable, the potential for improvement always exists.

The pope's almost certain to push back against this idea of improvement. Our universe is neither the result of a random lottery nor interchangeable with a host of others, as I've claimed. No, it's truly exceptional, designed specifically for us by its creator to offer us joy and beauty with every frame. Just witness a sunrise from a mountaintop, ponder the majesty of the land and the sweep of oceans beyond. You'll feel how perfect everything is, how lovingly for you it's been made.

To which I must respectfully offer a rebuttal similar to before. Landscapes form iteratively, whether driven by God or physics. They're the palimpsests left behind by successive geological ages. Mountains rise due to volcanic activity, rivers cut through land, valleys fill with water to form lakes. We've become accustomed to regarding the results of such input/output algorithms as beautiful, deriving joy from them. But such reactions are entirely learned, we aren't born that way.* We'd probably be wowed no matter what the equivalent of a sunrise was in any alternative universe we grew up in. So our feelings are unreliable indicators, they don't objectively prove the universe exceptional in any way.

Surely *life* does, though? I can see the pope's saved this trump card for the end. Evocative landscapes are one thing, but one can hardly argue that the presence of life-forms doesn't make our uni-

---

* Another learned reaction, which we mentioned in Chapter 26, is to recognize landscapes as "realistic" only if they have some underlying fractal structure.

verse unique. That this doesn't moreover establish Nature as God, not physics. Do I really want to claim life can be composed using mathematics? Am I arrogant enough to believe this is even within the purview of my subject?

The creation of life is undoubtedly the deepest gap separating those who equate Nature with God from those in the physics camp. I'm not talking so much about the question whether life-forms evolved iteratively to their present state (since, as proposed above, iteration could fit into any plan for creation, divine or not). The more irreconcilable split is about that first spark that ignited life. Was it miraculous, or can science explain it? What role could the mathematics we've developed in this book have played?

The basis of any scientific explanation for life does, indeed, come from the numbers—from the phenomenon of randomness. Count all the chance interactions that occur—not just on Earth but on every planet in every galaxy of any universe, and you keep increasing the odds. That the right combination of chemicals and catalysts and environmental conditions will somehow create the first living molecules, the first biological building blocks. With enough opportunities, you're likely to hit the life jackpot. (Keep in mind that "life" may need a broader definition if we're talking about different types of possible universes.)

But there's still the matter of that spark. How could a group of inanimate ingredients combine to suddenly attain such a dramatically enhanced state, that of being alive? What might mathematics have to say to explain such a boost?

Emergent phenomena

To answer this, let me take you back to the section on patterns, where we saw how a mollusk might create the Sierpinski triangle on its shell. The math there was very simple—an input/output scheme involving just three neighboring cells. And yet, by repeating this completely elementary rule, a pattern with much more profound complexity emerged. Similar rules can be set up to explain the spontaneous appearance of other recognizable phenomena—spots on a leopard, stripes on a zebra, human fingerprints, the complex patterns in which birds flock. For instance, the figure below is from a computer simulation that shows how patterns can form on the skin of an animal as it grows, starting from an initial random state of coloration and following fairly simple mathematics (using ideas originally proposed by Alan Turing).

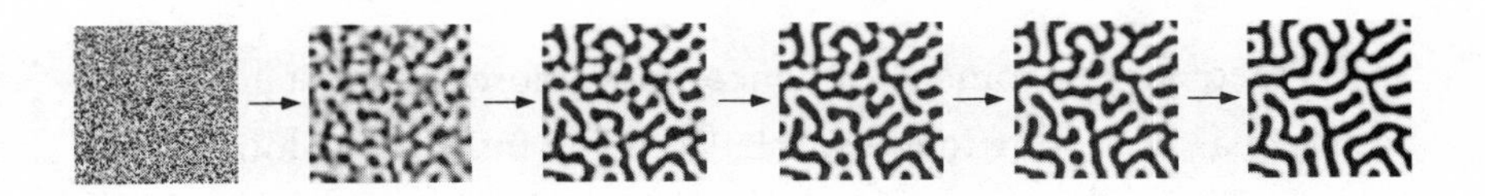

Here's the intriguing thing. The mollusk isn't consciously trying to generate a particular pattern. Nor is the leopard, zebra, birds, or we humans, when it comes to the ridges on our fingertips formed during gestation. In fact, the rules that create the ultimate design are entirely local—involving just a few cells adjacent to one another. In other words, the patterns are unpremeditated, self-arising—or, as referred to in complexity theory, "emergent." Contrast this with an artist who has a clear mental picture of the end result, and brings it to realization by applying calculated dabs of paint on different parts of the canvas.

We've encountered additional emergent phenomena in earlier chapters. Recall the particles herded into a knot in Chapter 26, that "magically" formed a sphere when they were instructed to each move forward one unit. The reason this sphere emerged was that the particles were facing random directions, not because of any intent or collusion on their part. Order emerged from randomness; all that was needed was a very simple instruction.

As another example, think of the intricate black-and-white fractals formed in Chapter 25 when individual points followed iterations like $z \rightarrow z^2 - 0.6$. Each point's fate was independent—it got colored black or white depending on whether it went off to infinity or not. Again, there was no collaboration between the points, and yet, a simple rule, followed individually, caused a surprisingly complicated pattern to emerge.

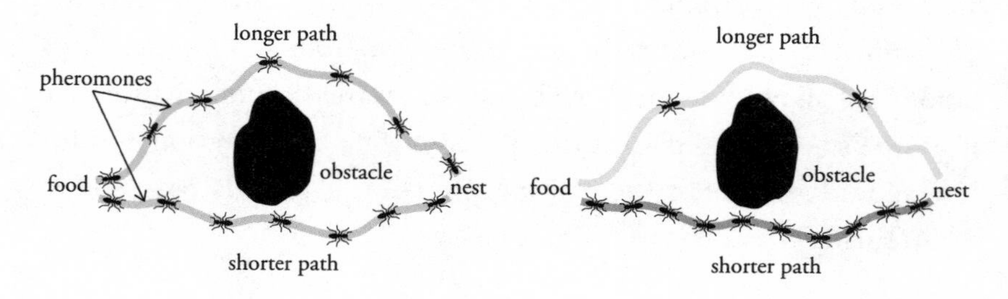

Such emergent phenomena can encompass more than just geometric patterns, as the following example from the insect kingdom shows. As ants randomly wander around, they might find several paths to the same food source. Other ants, sensing the pheromones dropped along the way by their predecessors, will follow. Shorter paths take less time, and will therefore be traversed by more ants, which drop more pheromones. Over time, these paths will consequently strengthen with increasing pheromones and ants, while longer routes will fade and die out. The ants will thereby have "learned" the shortest route to their food through this feedback loop, without anyone being in charge.*

Rather than a pattern, the phenomenon to emerge this time is "swarm intelligence," the ability of the group to act in a coordinated way as a single intelligent organism, even though each individual is following a simpleminded rule (dropping and tracking pheromones). Similar emergent intelligence can explain many seemingly complicated tasks insects perform, for example, constructing intricate bee-

* This is the example we promised earlier, about learning and feedback loops that can proceed autonomously, without any outside control. Perhaps Nature could devise similar automatic feedback rules to fine-tune gravity, physical constants, and so on, as she iterates her way to our universe.

hives and termite nests, or building miraculous army ant bridges through the air.*

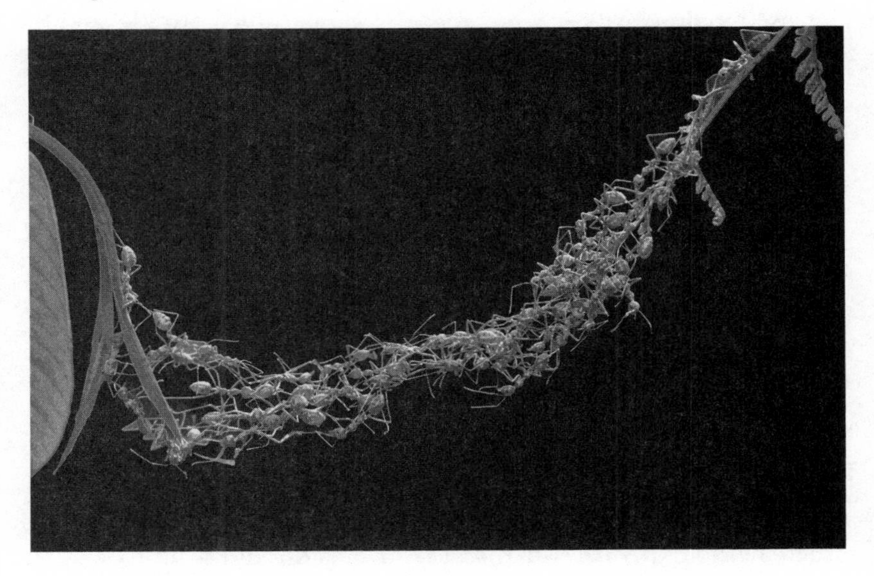

These examples bring us to a pivotal question. If intelligence can be an emergent phenomenon, could life be as well? Could "life" simply be a name we give to the complexity that emerges when the right constituent components, environmental conditions, and rules of interaction come together? Most scientific explanations of life center on this notion, though there is much work left to be done in creating a fully convincing model. What mathematics brings to the table is the insight that starting with very simple foundations, jumps in complexity *can* be attained, emergence *can* occur. Shell fractals and ant intelligence might appear magical, but these are just manifesta-

* The algorithm underlying ant bridges is also simple. As ants in a procession come to an air gap, they slow down, causing the faster ants behind them to clamber over. The elementary rule the ants follow is this: if you feel another ant's weight on you, freeze. More layers of ants keep climbing over, slowing at the gap, and freezing when ants climb over *them*. Each layer of bodies projects a little farther forward, with the result that eventually a chain of frozen ants bridges the gap.

tions of simple underlying mathematical rules. Life is similar, one might argue, no matter how miraculous its existence might seem. The increase in complexity it represents arises logically, naturally, and by statistical chance. A supernatural spark is not needed.

The twain

So what we have are two contrasting ways to view life: as an unpremeditated byproduct of input/output rules; and as something divinely bestowed, sacred. The same two perspectives extend to the creation of the universe at large. The driving distinction between them is *intent*. Yes, there might be differences regarding how Nature is supposed to do her building—instantaneously, or through an iterative process. But both sides would agree she's been known to use either method (*iterative* in the case of erosion and a host of other uncontroversial natural processes; *instantaneous* in, for example, the big bang). What really sets the two philosophies apart is whether Nature has a plan. Is the universe the result of volition? Has it been created for a reason? Or does it just exist? With no intrinsic purpose attached to it?

The pope, naturally, is on the "creation with intent" side of this divide. Just look at the universe, he'll say, and you will see it has both beauty and purpose to it. This is self-evident, not just what the Church teaches.

Others might feel differently. Intent, beauty, purpose, and so on, are what *we* read into it, what *we* imbue it with. The universe just exists.

This twain can never meet. Which is worrisome. Suppose I were to confess to being on the opposite side to the pope. Would this end our budding friendship?

Fortunately, mathematics comes to the rescue—by demonstrating how both views can be simultaneously accommodated. For, at its core, it harbors a similar dichotomy. Does mathematics exist because we create it, because it has a purpose? In other words, is it a product of our volition?

Or does mathematics simply exist? A logical system that self-generates, using only nothing as its basic ingredient, and proceeding all the way to infinity through numbers, patterns, formulas, shapes? One that will always be there, whether or not it's ever applied or understood or appreciated, whether or not God, physics, or the universe exists?

Both viewpoints have merit. Both have demonstrated, in the previous chapters, how applicable they are to the subject. One characterization may resonate with you more than the other, but there's no "proof" you can muster to validate your leaning. We all have to live with this duality.

In fact, we should treasure it. The provenance of the most perfect system known to humankind cannot be nailed down. Could this be the ultimate metaphor, the deepest insight math offers us? The idea that such duality between design and accident, intentionality and purposelessness, is not a question to be resolved one way or the other, but an essential feature of the way things are, an inextricable part of reality's DNA? We encountered something like this earlier with incompleteness—a limitation that at first glance may have disappointed, even suggested a defect. But when we stripped our own expectations away, incompleteness simply characterized an intrinsic aspect of what mathematics *is*. Now we've uncovered another evocative facet of the subject. The broader lesson it inspires is that we may never be able to determine whether our own existence is a result of purpose or randomness. All we can do is choose the perspective we're most comfortable with.

This, then, is how math speaks to us about our origins—an insight I promised at the beginning of this book. The pope and I have common ground on which to take our strolls again.

The adventure continues

Now that our time together, dear reader, is coming to an end, a key question arises. Did our thought experiment succeed or didn't it? In other words, were we able to establish that the universe can, indeed, be built using only math?

In my assessment, I'd have to say we did pretty well. Starting with the construction of empty space (which most origin narratives take for granted) we went on to create shapes, formulas, patterns, laws, and a way to implement them. One could nitpick that we didn't connect every dot or that we left too much to input/output processes. But remember, every alternative accounting, religious or scientific, has at least a few gaps. Our effort holds its own among them.

Moreover, the real motivation behind our thought experiment was to provide a tour of mathematics, with new insights into the subject. This is what we should use as the true measure of the experiment's success. Did we challenge fundamental beliefs? Were there Big Questions along the way? Out of the manifesto promised in Chapter 1, were all the boxes checked?

If so, let me invite you to continue with this thought experiment. To keep looking at the universe through the perspective you've gained, with every phenomenon and experience now imbued with math's clarifying radiance. Space swirls around you and unfurls under your feet, in invisible ways you may never before have imagined. Curves plunge along the tracks of hurtling roller coasters, and unspool lazily in gardens to form fruit and flower shapes. Fractals carve out intricately repeating coastlines and create spectacular current and cloud displays. Input/output processes cycle unseen everywhere, manifesting themselves all the way down to the whorls on your fingertips. Wherever you turn, the scent of infinity, evocative yet elusive, is always in the air. You can stay here as long as you want—the universe will always be changing, so the experiment can never end.

In fact, Nature's just launched a host of new input/output processes to run on autopilot while she takes a break (Day 7, she's heard, is supposed to be a day of rest). As you watch these budding cycles, you think how math will determine their fate. Some might iterate their way to tiny life-forms, some to enormous galaxies, most will just fade away. You try to latch on to what mathematics *is*—this exuberant, omnipresent, omnipotent force obeyed by everyone and everything—believers, rationalists, even the inanimate. A force that forever enthralls, not just through the answers it gives but also through the new mysteries it poses. You sit there, under the stars and the sky, feeling the ground beneath you and the breeze on your face.

ACKNOWLEDGMENTS

WRITING A BOOK, MUCH LIKE DESIGNING A UNIVERSE, IS AN iterative procedure. You create a series of drafts and the manuscript (you hope) improves with each step. Fortunately, I've received a lot of help through the process. Thanks, first and foremost, to four people who've seen me through the publication of all four of my books—my agent, Nicole Aragi; my editor, Jill Bialosky; my grad school buddy Nancy Pfenning; and my husband, Larry Cole. Without your support, encouragement, and critiquing, this "Big Bang" might never have made it to completion (thanks, Larry, for your photography and Photoshopping as well!).

Getting feedback from readers has been particularly crucial for this project. The comments provided by Ben Bowden, Jessica Yao, and Drew Elizabeth Weitman (all of W. W. Norton) have been truly invaluable. Many thanks too, Drew, for all the time, effort, and care you've put into shepherding this book through production.

Other readers to whom I am indebted are Shirley Cole, Mia Pellegrini, Kathrin Luksch, Tom Ramsey, John Feffer, John Zweck, Sue Brenner, and Karen Kumm. Thanks also to Daina Taimina, Ser Tan Peow, and Bill Goldman for conversations related to the section on geometry (and to Lauren Schick, my crochet consultant!); Markos Georganopoulos for his comments on the physics; Frank Pfenning for a computer science perspective; and John Allen Paulos for overall reactions to the manuscript. A special thanks to Steven Strogatz for so generously sharing his insights into writing mathematics books for general audiences.

I'm indebted to the University of Maryland, Baltimore County community for its longtime support of my writing activities. Thanks in particular to Kathleen Hoffman, Thomas Mathew, and other members of my department with whom I've consulted, to UMBC students Kevin Kauffman and Lana Hill for their detailed comments on early chapters, to Visual Arts faculty members Lee Boot and Kelley Bell for video explorations of some of this material, and to President Freeman Hrabowski for taking such a personal interest in this project. Many thanks to Michele Osherow from the Department of English, my partner in crime on a humanities seminar (and subsequent play, *The Mathematics of Being Human*), which helped set some of the topics in this book in place. Above all, thanks (or perhaps apologies) to students in my two classes from Fall 2018, Introduction to Contemporary Mathematics and The Godfather of Numbers, in which I used an earlier draft of this book as the text. This version is much improved because of your feedback!

I'm grateful to Dae In Chung for doing such a brilliant job composing over two hundred illustrations in a tight time frame, and to Jose Villarrubia for his deft and evocative hand drawings. Shelley Husband from Australia, Ted Burke from Ireland, and Sebastian Valette from France all created illustrations especially for this book—efforts that are much appreciated. Edna St. Vincent Millay's poem, "Euclid alone has looked on beauty bare," from which I have quoted lines, is now in the public domain: I would like to acknowledge Holly Peppe, literary executor of the Millay Society, for informing me of this.

Thanks to Julia Druskin, Dassi Zeidel, and the rest of the production team at W. W. Norton for bringing such a visually striking book together from so many disparate pieces, and to Rachel Salzman and Steve Colca for all their promotion and marketing efforts. A special shoutout to Ingsu Liu for designing the gorgeous cover. I'd also like to thank Caspian Dennis (from Abner Stein), Natalie Bellos, and Rowan Yapp for making the Bloomsbury UK edition possible—complete with its own arresting cover designed by Mike Butcher. A heartfelt thank you to Honor Jones, who first approached me with the idea of writing the opinion piece on mathematics for the *New York Times* from which this book developed.

Finally, I'm grateful to the Sloan Foundation for supporting me through a generous book grant, and to the Rockefeller Foundation for a glorious residency at the Bellagio Center, Italy, where part of this book was written.

NOTES

Introduction

1 ***my opinion piece:*** This is available at https://www.nytimes.com/2013/09/16/opinion/how-to-fall-in-love-with-math.html.

2 ***PowerPoint talk:*** "Taming Infinity" is available at https://www.youtube.com/watch?v=kBS_cNHvnBE.

3 ***religion had handily won:*** I also learned this lesson at the 2013 Jaipur Literature Festival, when my event on the Indian mathematician Ramanujan drew about ten people (in a space meant for perhaps a thousand). It turned out the Dalai Lama was speaking next door at the same time. His event got so crowded, though, that people started wandering over to mine. By the end, thanks to him, we also had a packed crowd.

3 ***little media attention:*** The quote is reported in the article "Math and the Media: A Disconnect, and a Few Fixes, Emerge in San Diego Session" by Sara Robinson, SIAM News 34, no. 8 (October 2001), https://archive.siam.org/news/news.php?id=577.

9 ***thought experiments:*** "Schrödinger's cat" is perhaps the most well-known physics thought experiment. A cat in a closed box exists in a state of being both dead and alive, since its fate is tied to the decay of a radioactive substance in a similarly superimposed state. The cat's fate is determined only when the box is opened, illustrating a fundamental paradox in quantum mechanics.

In mathematics, the thought-experiment tradition is less developed, possibly because *every* abstract argument could be considered to be one. Narrative riddles like Xeno's paradoxes—based on the apparent impossibility of summing an infinite series—probably come closest to the ones in physics. Our thought experiment here is more expansive than many traditional ones, and straddles both physics and mathematics. For an interesting treatise on the philosophy behind thought experiments, see J. R. Brown and Y. Fehige, "Thought Experiments," *The Stanford Encyclopedia of Philosophy* (Winter 2019 Edition), Edward N. Zalta (ed.), https://plato.stanford.edu/archives/win2019/entries/thought-experiment/.

11 ***more of a story:*** The usual way to bring out the story aspect of math is to trace its historical development. However, mathematical discoveries didn't historically emerge in a straightforward sequence, so our narrative here allows for a more logical development.

11 ***my most treasured potential reader:*** Although the pope may not be into cricket (which hasn't been popular in Argentina for decades), he's known to be a soccer fan. He confesses to not being a good player—while playing, his nickname was "Stiff Legs" and he was always relegated to goalie position (see the NBC News video "Pope Francis Tells Kids He Wasn't Good At Soccer Growing Up," https://www.youtube.com/watch?v=C2APV68x7gA). But this did not make him lose interest in the game. Which jibes perfectly with my book's message of not having to be good at something—be it soccer or mathematics—in order to appreciate it.

Day 1: ARITHMETIC

16 ***quest for numbers:*** Research shows that animals have some innate sense of number, so humans probably inherited that. However, refining this to an abstract mathematical concept was an enormous advance.

16 ***six-sided cube-like chunks:*** See J. Sokol, "Scientists Uncover the Universal Geometry of Geology," *Quanta Magazine*, November 19, 2020, https://www.quantamagazine.org/geometry-reveals-how-the-world-is-assembled-from-cubes-20201119/.

19 ***the Hindus invented zero:*** Although other cultures used related concepts, the Hindus are usually credited with being the first to recognize zero as a number, possibly due to the essential role played by the void in their religion. This was in contrast to the Aristotelian philosophy of the Greeks, which rejected the void, even feared it. (See, for example, Charles Seife, *Zero: The Biography of a Dangerous Idea* [New York: Penguin, 2000]. Also, here's a fun *New York Times* opinion piece I wrote on this topic, from October 7, 2017, entitled "Who Invented 'Zero'?": https://www.nytimes.com/2017/10/07/opinion/sunday/who-invented-zero.html.)

20 ***the number of dots on it:*** There are alternative ways to define whole numbers. For instance, the number n could be defined as the collection of all sets containing n elements. Another option is to define 1 as the set containing 0, followed by 2 as the set containing 1, and so on (compare this to what we've used, where 2 is the set containing both 0 and 1).

23 ***Numesthesia:*** David Kessler, who wrote this play, performed it at the 2017 Capital Fringe Festival in Washington, DC. See V. Hallett, "When This Man Sees Numbers, He Sees Them as People," *Washington Post*, June 24, 2017, https://www.washingtonpost.com/national/health-science/when-this-man-sees-numbers-he-sees-them-as-people/2017/06/23/0afad744-55ff-11e7-b38e-35fd8e0c288f_story.html.

23 ***on genders and pronouns:*** I've assigned genders to numbers as a nod to Lili's experience, but these can be ignored or mentally changed without making a difference to the narrative. On Day 3, in accordance with most religious and mythological traditions, Nature is introduced as female—again, this gender choice does not play any significant role in the story. I haven't attached gendered pronouns to the general concept of God, with two exceptions. First, I have the pope use masculine pronouns for God in keeping with current custom in the Catholic Church. Second, to try and balance this out, I've used feminine pronouns for such a divine entity at the start of Chapter 27.

24 ***if one defines sets "naively":*** Perhaps the most well-known problem is *Russell's paradox*. We saw in Chapter 1 that a set can contain other sets as elements (e.g., the set for 1 contains the empty set as an element). So here's the question that the

mathematician Bertrand Russell posed: Could a set contain itself as an element? Certainly, there's nothing in the usual definition of sets to prevent this.

Russell then pointed out that sets could be of two kinds—those that contain themselves as an element, and those that do not. For instance, "the set of all ideas" is also an idea, so it contains itself as an element. Similarly, "the set of all sets" is a set, and "the set of non-umbrellas" is not an umbrella—so both these sets also belong to themselves as elements.

Suppose now one considers "the set of all sets that do not contain themselves"—call this set A. Does A contain itself as an element or not?

Let's say the answer is yes, i.e., A *is* an element of itself. By the definition of A, this means A is a set that does not contain itself. But what this implies is A would *not* be an element of itself!

So let's say the answer is no, A is *not* an element of itself. But this is the exact characterization of sets included in A. So A *would* be an element of itself!

We see therefore that either way, we get to a contradiction. There's no way out, really—and the cause is simply this: we were too casual in defining a set as "a collection of objects." This is what Russell showed with his paradox: to be free of logical inconsistency, one cannot define a set so naively.

After Russell's paradox became known, logicians worked furiously to repair set theory. They found that the only way to do so was to carefully make a list of basic assumptions called *axioms* from which everything could be built up without contradictions. This allowed them to design axioms in such a way that Russell's self-containing sets were excluded from consideration. Further axioms addressed other issues that had been taken for granted so far. For instance, rather than simply invoking an empty set, they added a basic axiom that *assumed* its existence, or a set of alternative axioms, from which the empty set's existence could be *proved*. Similarly, the justification that our "Big Bang" construction of whole numbers could go on indefinitely was provided by an "axiom of infinity" that explicitly declared such an infinite process permissible.

The most commonly used list of foundational axioms for set theory is the one formulated by Zermelo and Fraenkel, which leads to the so-called ZFC system.

26 ***G. H. Hardy:*** In his 1940 work, "A Mathematician's Apology," Hardy rousingly extols beauty in math, while disdaining "useful" math, which he equates with those parts of the subject that he considers "dull" and "ugly."

27 ***the "game" of* addition:** Although we didn't do so in the main narrative, we can formally define addition in terms of our defining sets. The gist is as follows. For any natural number n, let $S(n)$ denote its successor $n + 1$ (in terms of our set-based definition, this is just the set $n \cup \{n\}$). Then we define addition recursively by the rules

$$a + 0 = a, \; a + S(b) = S(a + b).$$

Once the above rules are clear, you will see that, for example,

$$1 + 1 = 1 + S(0) = S(1 + 0) = S(1) = 2.$$

Using the above notation, one can also define multiplication recursively by the rules

$$a \times 0 = 0, \; a \times S(b) = (a \times b) + a.$$

29 ***to represent* any *number:*** It was only in 1202 that this convention was significantly introduced to the West, by Fibonacci, in his book *Liber abaci*. He noted,

at the start of the first chapter, that *any* number could be written with the nine Indian digits 1 through 9, and the sign 0. This was an enormous breakthrough, representing centuries of experimentation and thought.

35 ***Textbooks renounced their use:*** For example, the 1796 textbook *The Principles of Algebra* by William Frend (published by J. Davis) contains this broadside on page x: "To attempt to take . . . away from a number less than itself is ridiculous. Yet this is attempted by algebraists, who talk of a number less than nothing, of multiplying a negative number into a negative number, and thus producing a positive number."

36 ***self-contained under +, ×, and −:*** The operational rules we have for naturals can be rigorously extended to the larger set of integers to demonstrate this. However, this is quite technical, so we omit it here. Looking ahead, similarly rigorous rules can be defined over larger sets of numbers (rationals, reals, complexes) to show that these sets are self-contained under various arithmetic operations. See the endnote for page 61 ("a set that's self-contained"), below.

37 ***"broken numbers":*** Although Nicolas Chuquet introduced this term in French, it is also found in books in English published well into the seventeenth century; see G. Flegg, C. Hay, and B. Moss, eds., *Nicolas Chuquet, Renaissance Mathematician* (Dordrecht, Holland: D. Reidel, 1985), 43.

42 ***introduce pi at this stage:*** Pi can be written as the sum of an infinite series,

$$\pi = 4\left(1 - \tfrac{1}{3} + \tfrac{1}{5} - \tfrac{1}{7} + \tfrac{1}{9} - \ldots,\right).$$

a formula attributed to Leibniz, but first discovered in the fourteenth century by the Indian mathematician Madhava. Notice this has nothing to do with circles; it involves only the arithmetic operations of +, −, ÷ that we've already defined. So imagine the numbers playing a game where they add together successive terms in this series, giving them answers that get closer and closer to pi. They can calculate pi to any accuracy by taking enough terms, and even calculate it *exactly* if we imagine them being able to complete the game. This infinite-series "game" for pi is the simplest, but others also exist.

47 ***unrelated to any rational:*** Irrational numbers of the first type (like the square root of 2 or the cube root of 3) are called *algebraic*. More technically, these numbers can be expressed as the roots of a polynomial equation with integer coefficients. The remaining irrationals, like pi, are called *transcendental*. Even though these form the bulk of the irrationals, they may not be as well known. One example is the number $e = 2.718\ldots$, familiar to those who've taken calculus, as the base of the natural logarithm.

48 ***were they to use forty digits:*** NASA's use of only fifteen digits for pi is mentioned, in the words of Marc Rayman, a mission manager, on their Jet Propulsion Lab website in a March 16, 2016, post, "How Many Decimals of Pi Do We Really Need?": https://www.jpl.nasa.gov/edu/news/2016/3/16/how-many-decimals-of-pi-do-we-really-need/.

50 ***"The Library of Babel":*** A neat computer simulation of Borges's library is available at the Library of Babel website, https://libraryofbabel.info/. Using this website, you can locate any passage of up to 3,200 characters in the virtual library.

51 ***every possible string of digits:*** The technical name given to numbers that have each string of n digits appearing in their decimal expansion with about equal frequency is "normal."

51 ***312 occurs 1,999,464 times:*** Thanks to James Taylor for performing this search, as well as searches of 312 in the expansions of $\sqrt{2}$ and the Golden Ratio. He pro-

vided me not only the frequency but also the exact locations of every occurrence in these expansions!

51 ***trillion digits of pi:*** Based on data from the Wolfram Mathworld website, maintained by Wolfram Research: https://mathworld.wolfram.com/PiDigits.html.

53 ***what constitutes "true" randomness:*** Random-number generators are of two types. Those based on numerical algorithms are called "pseudo" random-number generators, because the fact that they are based on a mathematical algorithm means they can't be truly random. (There are different definitions for what qualifies as "pseudorandom," which further complicates the classification of algorithms.) So-called "true" random-number generators are based on some unpredictable physical phenomenon such as radioactive decay or atmospheric noise. However, the unpredictability could be due to the limitations of current physics theories, so absolute randomness is still hard to pin down. See Mads Haar, "Introduction to Randomness and Random Numbers" on Random.org, https://www.random.org/randomness/.

61 ***the end of an evolutionary chain:*** Actually, there are number systems even larger and more general than the complexes (e.g., the so-called quaternions and octonions). However, these are far less frequently encountered than the complexes.

61 ***a set that's self-contained:*** Two complex numbers $x = a + ib$ and $y = c + id$ can be added as follows: $x + y = (a + ib) + (c + id) = (a + b) + i(c + d)$. This is clearly a complex number, with real part $(a + b)$ and imaginary part $(c + d)$, so the complexes are self-contained under addition. Subtraction is similar. Multiplication, division, and root-taking formulas are more involved, but you can easily look them up if interested. In all cases, you end up with another complex number as the answer.

63 ***what we've accomplished:*** The Prussian mathematician Leopold Kronecker famously declared, "God made the integers, all else is the work of man." However, as we've shown, humans can make the integers as well!

Day 2: GEOMETRY

76 ***extend this ordering game:*** This was explained for positive rationals earlier, but the decimal expansions of positive irrationals could also be included in the ordering. For negative numbers a and b, we could then just order them as $a < b$ when $-b < -a$.

79 ***little math miracles:*** One could argue the pope is correct, that axioms *are* like little math miracles. This goes back to the fact that there can never be absolute *creatio ex nihilo*. Axioms are the starting assumptions of math, from which the rest of the subject follows by logical deduction.

To be useful, the system of axioms one starts with needs to be *consistent*, i.e., they should not lead to logical contradictions. Also, a desirable property is that the axioms in the system be *independent*, i.e., you shouldn't be able to derive one of them from the rest. Needless to say, they should have a high degree of plausibility, to the extent that they should be generally regarded as "self-evident."

Building up mathematics axiomatically ensures that each proposition you prove is definitively true if you begin with your starting assumptions. If you end up proving a result that is dubious or outlandish, you can proceed backward and figure out how your starting axioms may be modified to exclude this result.

The fact that axioms need to be plausible, consistent, independent, and the precursors of all logical deductions to come makes them more coherent, dependable, and easily acceptable than miracles.

But there's an inescapable reason why axioms are *not* miracles. A miracle is supposed to be absolutely true, while an axiom can never be proved, only assumed. The world of math you end up with, starting with a set of axioms, does not have any absolute physical reality, though depending on how well you choose your axioms, it could be a good *model* for such reality.

84 ***check out the endnotes:*** First of all, note that our two-point method gives *every possible* line on the plane we end up constructing, except for those parallel to the first or second line, and those passing through the point of intersection of these two lines. Indeed, take any two points on the plane and draw the line l through them. Then, unless l is parallel to the first line, it will intersect the first line at some point A. Similarly, unless it is parallel to the second line, it will intersect it at some point B. Eliminate the case when A = B, i.e., when l passes through the point of intersection of the first two lines. Then, joining A and B, we get l via our method.

This shows that every point on the plane is, indeed, generated by our method. The point of intersection is generated at the start, while every other point P lies on an infinite number of lines l that pass through P and are not parallel to either the first or second line.

Now, here's a method with much less duplication. Instead of joining every pair of points, you need to join just the new point to successive points on the original line. This will give you the plane, formed by the lines radiating from the central point they have in common. The only problem is that the central horizontal diameter h of the circle shown will be missing. To address this, take any point that's on neither h nor the original line, and join it to every point on the original line. Then h will get filled in.

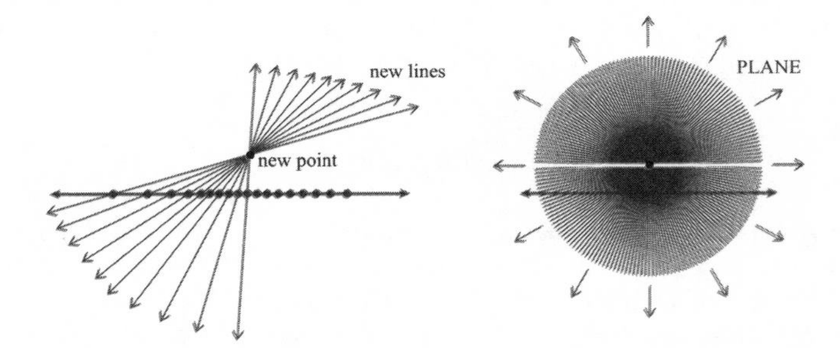

95 ***different points taken as center:*** Although the numbers discover this through play, the proposition that one can draw a circle with any center and any radius on an infinite plane can, strictly speaking, be justified only through an axiom. In fact, this is precisely Euclid's third axiom (after the previous two about straight lines on a plane).

Let us here also mention Euclid's fourth axiom, which says that all right angles on a plane are equal. This means that if two lines intersect such that adjacent angles are equal (i.e., they are "right"), then such angles will match those obtained from the intersection of any other two lines.

Euclid's third and fourth axioms help characterize a plane by telling us that its geometric nature is uniform all over its (infinite) expanse. Were we not on a plane, these axioms might not hold. For instance, suppose we were on the surface of a sphere instead. Then one could never draw a circular segments larger in radius than the radius of the sphere on this surface. Similarly, if we were on the surface of a cone (or on any surface that had a cone-like vertex on it), then due to this nonuniformity, "right" angles may not always be equal. The fourth axiom also tells us that our "straight line" cannot have any corners. See R. Rucker, *Geometry, Relativity and the Fourth Dimension* (Mineola, NY: Dover, 1977), 19–20; and also the end of Chapter 4 of D. Henderson and D. Taimina, *Experiencing Geometry: Euclidean and Non-Euclidean with History*, 4th ed. (Project Euclid, 2020), https://doi.org/10.3792/euclid/9781429799850.

96 ***these segments intersect:*** This construction of an equilateral triangle is the first of Euclid's propositions (i.e., theorems). The proof he gives is simple, but has gaps—chief among them being how one can *prove* that the two circular segments will intersect. It turns out one needs additional axioms to plug in such gaps (an augmented list of twenty axioms was formulated by the German mathematician David Hilbert in 1899). Although some other proofs by Euclid also have gaps, what is remarkable is that the statements of all his propositions are correct.

Notice that if we draw entire circles, instead of just circular segments, we will get two points of intersection, and hence two different equilateral triangles.

103 ***construct 3-D space:*** To construct space using the two-point axiom, first connect the new point with *any* point on the complex plane. This gives a new line. Now join every possible pair of points, where one point is on the new line and the other is on the plane. Each such connection results in a line, and the totality of all points on these lines will give 3-D space.

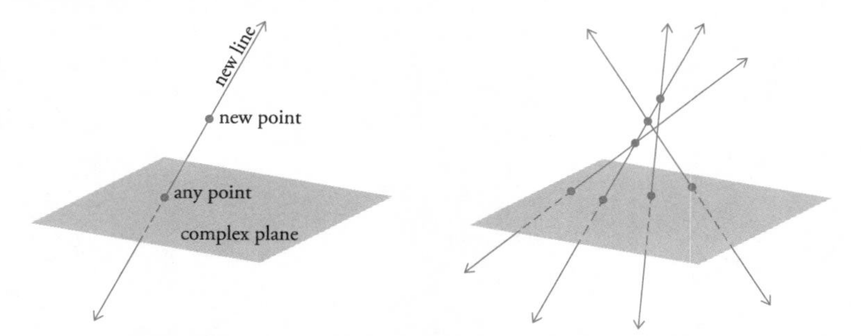

103 ***we're set to go:*** Constructing space via the parallel axiom again presents a chicken-and-egg problem (see the end of Chapter 8), since we call two planes parallel if *they are in the same 3-D space* and never meet. So the best way to think of this is that the two-point axiom is first used to construct 3-D space, as explained above. You can then find the line through the extra point that's perpendicular to the complex plane (though the construction also works if this line is tilted instead of being perpendicular). For each point on this line, the parallel axiom now guarantees a plane passing through that point which is parallel to the complex plane.

106 ***evicting the complexes:*** Historically, the idea of using complex numbers to represent points on the plane can be traced back to a 1799 Dutch paper by one Caspar Wessel. This was well after the use of ordered pairs to do so was proposed by Descartes and others who followed (starting with a paper in 1637). So complexes were never "evicted" in favor of ordered pairs as our narrative suggests. However, complexes never became popular for day-to-day usage on the plane as "Cartesian" coordinates did, and one might speculate that the fact that they could not be extended easily to 3-D may have played a role. Currently, the complex plane is used in select situations, for instance when one wants to "multiply" points together (see Chapter 25).

110 ***A version of the axiom will still hold:*** Let's first note that the two-point axiom for the plane is stated more completely as "Given any two points on a plane, we can construct an endless line on that plane that passes through them" (though see note 1 at the end of this entry). The analog for a sphere is therefore "Given any two points on a sphere, we can construct an endless line (great circle) on that sphere that passes through them." This axiom for the sphere is true, though we must interpret "endless" not as *infinite* but as *unending*, i.e., wrapping around to form a continuous loop.

Now let's list the steps by which the procedure in Chapter 8 generates a particular plane.

Step 1: Take two points on the plane and join them by a line.

Step 2: Take a new, third point on the plane that does not lie on this line. Join this third point to an arbitrary point on the line to create a new line.

Step 3: Start drawing lines between pairs of points on the original and new lines. This multitude of lines will, in the limit, yield the plane.

Precisely the same procedure works for the case of a sphere. All you need do is replace "plane" with "sphere" in the above list, and interpret "line" as "great circle." The diagrams below pictorially explain the analogy.

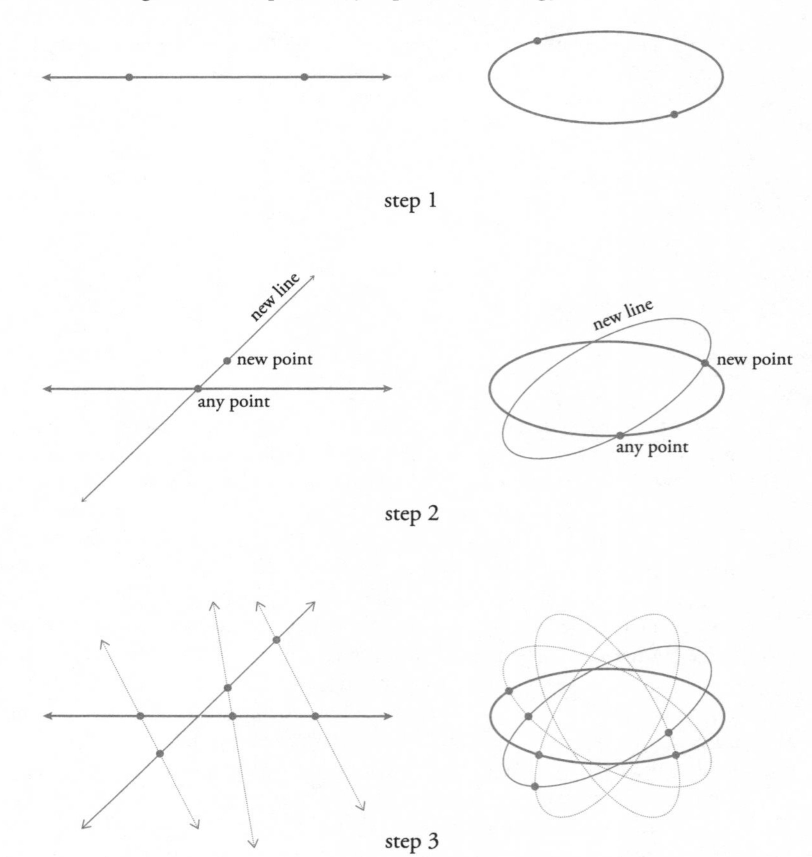

NOTE 1: There is one significant difference between the constructions for the plane and for the sphere. Suppose we state the two-point axiom as "Given any two points, we can construct an endless 'line' that passes through them." For the planar case, step 1 will then give a fixed line. If we're prepared to end up with an *arbitrary* plane (as opposed to a *particular* one), we can simply choose *any* new point in step 2 (one that doesn't lie on the line from step 1). This will give us *some* plane once we complete step 3.

Now consider the spherical case, where "line" is interpreted as "great circle." Step 1 gives us a whole bunch of great circles to choose from (with different radii and centers), since two points do not determine a unique circle. If we select a particular great circle at this step, then step 2 will work *only* provided the new point lies on the sphere corresponding to this great circle. If we want the construction to work for other cases, we need to coordinate the selection made in step 1 with the choice of the extra point in step 2, so that everything is compatible. (This means going back and taking a different great circle between the first two points if necessary.)

NOTE 2: As explained in the endnote for page 84 ("check out the endnotes"), our two-point axiom method gives almost all lines that lie on the plane, so it is very close to the method of simply drawing all these lines to "get the geometry" that we carried out in the main text. For the sphere, the above two-point axiom method will again give almost all the great circles that lie on the plane, except for the ones that pass through the intersection of the first two great circles shown in step 2.

NOTE 3: One can also build the hyperbolic plane described in Chapter 13 by using the above method of construction based on this two-point axiom (you'd just use the corresponding analog of "straight lines" instead).

114 ***crocheting websites:*** The images here are from https://spincushions.com, but see also, for example, https://www.interweave.com.

118 ***on corals and crocheting:*** Daina Taimina's book *Crocheting Adventures with Hyperbolic Planes* (Boca Raton, FL: CRC Press, 2018) provides several other ways that the hyperbolic plane can come alive through tangible objects. Her main message extends far beyond the hyperbolic plane, though—she points out that tactile exploration is an essential way to experience and learn mathematics, rather than relying on just the standard method of abstract thinking. It would be well worth it to explore this idea in terms of not just school-level but also university-level math.

Taimina's work has helped inspire a massive worldwide project called the Crochet Coral Reef (https://crochetcoralreef.org/about/theproject/), developed by Christine and Margaret Wertheim. This project, which has thousands of (predominantly female) contributors, combines hyperbolic geometry and art to focus attention on environmental concerns, particularly the dangers climate change poses to coral reefs.

118 ***sea slugs, flatworms, and nudibranchs:*** See, for example, "Mathematics of Sea Slug Movement Points to Future Robots," Eurekalert, March 7, 2019, https://www.eurekalert.org/pub_releases/2019-03/aps-mos022219.php.

119 ***texts on geometry:*** For instance, see chapter 5 of Henderson and Taimina, *Experiencing Geometry*.

120 ***fifth and final axiom:*** Euclid's stated his fifth axiom (he called them "postulates") as, "If a straight line intersects two straight lines to make the interior angles on one side less than two right angles, then the two straight lines, if extended indefinitely, meet on that side."

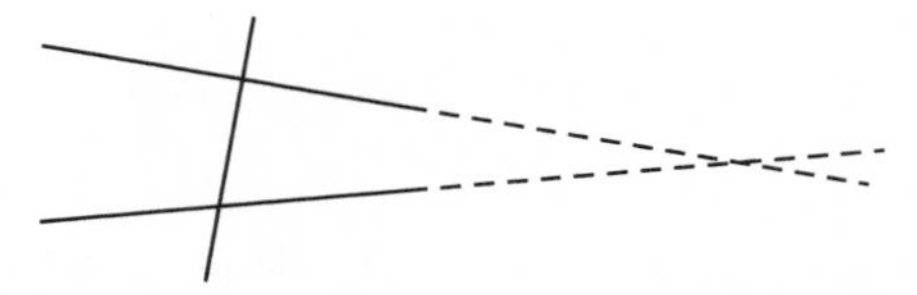

The "parallel axiom" we use is often called Playfair's axiom (the name is erroneous; see chapter 10 of Henderson and Taimina, *Experiencing Geometry*). Our version is equivalent to Euclid's version on the plane and hyperbolic plane, but is *not* equivalent on the sphere. That's because any two "straight lines" on the sphere intersect, so Euclid's version is trivially true, while our parallel axiom, as we've seen, is false.

120 ***a perfectly valid alternative:*** Wouldn't spherical geometry, which has been well known since ancient times, have already demonstrated the existence of "non-Euclidean" geometries? Apparently not, since this geometry was somehow never regarded as an alternative geometry in its own right. Moreover, unlike the parallel axiom we have used, Euclid's version was trivially true on the sphere, so there was no contradiction. Later on, however, Riemann developed so-called *elliptic geometry*, which generalized the case of spherical geometry and was recognized as another valid "non-Euclidean" alternative to plane geometry.

123 ***a 4-D setting:*** This isn't technically necessary, though. For instance, a sphere or hyperbolic plane could very well exist just by itself, without the benefit of a surrounding 3-D space. But imagining the 3-D surrounding makes it easier for us to picture these geometries. So, also, having a 4-D space in which a curved 3-D space is embedded may make it easier for us to grapple with the concept.

129 ***Gauss:*** The story of Gauss testing the curvature of space is almost surely apocryphal. See Arthur I. Miller, "The Myth of Gauss' Experiment on the Euclidean Nature of Physical Space," *Isis* 63, no. 3 (1972): 345–48, www.jstor.org/stable/229274.

129 ***uncertainty error of these experiments:*** The 0.4 percent error figure comes from "Will the Universe Expand Forever?" on the NASA website (updated January 24, 2014), https://map.gsfc.nasa.gov/universe/uni_shape.html.

129 ***positive curvature:*** For a sphere, the curvature is defined to be $1/r$, where r is the sphere's radius. So the curvature is always positive, and the larger a sphere, the smaller the value of the curvature. A flat plane can be thought of as the limit of a sphere as r approaches infinity: its curvature is therefore 0. A hyperbolic plane is said to have constant *negative* curvature. An explanation of this concept, together with some different types of curvature, can be found, for example, in Henderson and Taimina, *Experiencing Geometry*.

131 ***a fourth space dimension:*** See, for instance, Fraser Cain, "Finding a Fourth Dimension," Universe Today, May 30, 2006, https://www.universetoday.com/8209/finding-a-fourth-dimension/; and Zeeya Merali, "Theoretical Physics: The Origins of Space and Time," *Nature* 500 (2013): 516–19, https://doi.org/10.1038/500516a. Interestingly, some quantum experiments (see, for example, David Nield, "Experiments Show the Effects of a Fourth Spatial Dimension," Sciencealert.com, January 6, 2018, https://www.sciencealert.com/experiments-show-dramatic-effects-of-fourth-spatial-dimension) have hinted at an observable fourth spatial dimension.

Day 3: ALGEBRA

154 ***strings leading up to invisible kites:*** One thing to note is that if one extends the graph to the negative x axis, then a difference between such "strings" arises. For even

powers of x, the strings are just reflections across the y axis (like the graph of $y = x^2$). For odd powers, this reflection gets further reflected across the negative x axis.

164 ***change the center:*** Note that this is just Euclid's third axiom being illustrated for the plane, i.e., that a circle can be drawn with arbitrary center and radius.

165 ***essentially eliminate all corners:*** Remember that a cubic has four knobs to adjust it, compared to a linear's two. This allows cubic segments to be fine-tuned so that they join together more smoothly (at each nodal point, you can now specify not only the x and y coordinates but also the slope of the curve you're trying to generate). So there are no inconvenient kinks at such points.

Even with just piecewise linears, kinks will subside as you increase the number of points, since such an increase automatically makes the angle at each kink closer to 180°. This improvement is seen below when 16 and 32 points are used (instead of 8) to generate the heart shape. With 32 points, the outline is very close to the original.

169 ***concurs with Pythagoras or not:*** See Francis E. Su et al., "Spherical Pythagorean Theorem," Math Fun Facts, https://math.hmc.edu/funfacts/spherical-pythagorean-theorem/. See also D. Velian, "The 2500-Year-Old Pythagorean Theorem," *Mathematics Magazine* 73 (2000): 259–72.

169 **Squaring the circle:** This is the name of a classical problem from antiquity: to draw a square, using only straightedge and compass, that is exactly equal in area to that of a given circle. Once pi was shown to be a transcendental number (in 1882), this problem was proved to be unsolvable. Our section doesn't delve into this problem, just borrows its name.

170 ***Washington, DC, metro system:*** See "How Are Metrorail Fares Calculated?," Planitmetro.com, November 15, 2012, https://planitmetro.com/2012/11/15/how-are-metrorail-fares-calculated/.

172 ***other formulas for distance:*** For instance, one can define the distance of the point (x,y) from the origin by $d = \sqrt[r]{|x|^r + |y|^r}$. Then the unit "circle," given by all points for which $d = 1$, is as shown for different values of r. (The case $r = \infty$ represents the formula $d = \max\{|x|,|y|\}$.)

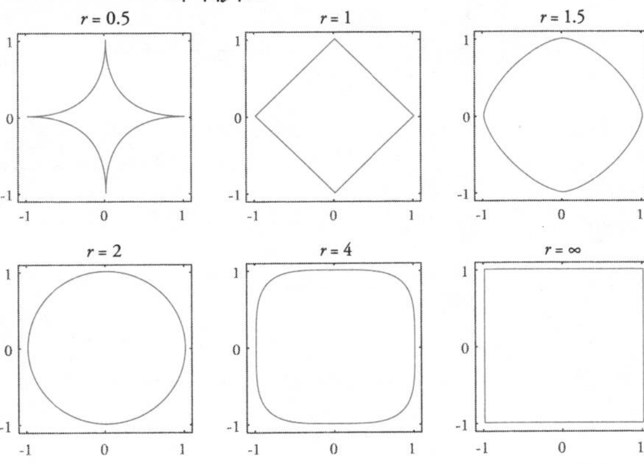

Day 4: PATTERNS

187 ***such symmetries:*** The mathematics of symmetry is usually expressed in terms of *group theory*, in the sense that the set of all transformations that leave an object invariant form a so-called "symmetry group," and the symmetry of an object can be classified in terms of such groups. See, for example, Mario Livio, *The Equation That Couldn't Be Solved: How Mathematical Genius Developed the Language of Symmetry* (New York: Simon & Schuster, 2005).

191 ***whittle down her work:*** There are many other reasons Nature is drawn to symmetry. See, for example, Livio, *The Equation That Couldn't Be Solved.*

201 ***close adherence to exponential growth:*** According to online analysis by Jacek Chmiel in "Data Science Perspective on COVID-19: A Real Life Example," Avenga.com, April 15, 2020, the data can be modeled more closely by a cubic polynomial. A similar cubic dependency was also found to better model the start of the AIDS epidemic (compared to an exponential model); see M. J. Harrison, "The Cubic Growth of AIDS Cases: General Dependence on Early Infection Rates and Distribution of Times to Appearance of Clinical Symptoms," *Journal of Mathematical Biology* 27 (1989): 523–35, https://doi.org/10.1007/BF00288432.

201 ***why it's called* exponential *growth:*** Let q be the growth factor, so that the population in each time period follows the rule $x \rightarrow qx$. Then it may be shown that if one starts with initial population x_0, the population after n time periods will be $x_0 q^n$. "Exponential" refers to the fact that n occurs in the *exponent* of the formula. As n increases, this growth is much faster than that modeled by powers of n, such as n^2 (quadratic), n^3 (cubic), etc. (The latter kind of growth is termed *algebraic.*) "Exponential" is often used to denote any type of fast growth—usage that can be distressing to mathematicians, as I commented in an opinion piece, "Stop Saying 'Exponential.' Sincerely, a Math Nerd," *New York Times*, March 4, 2019, https://www.nytimes.com/2019/03/04/opinion/exponential-language-math.html.

203 ***Euclid was the first:***

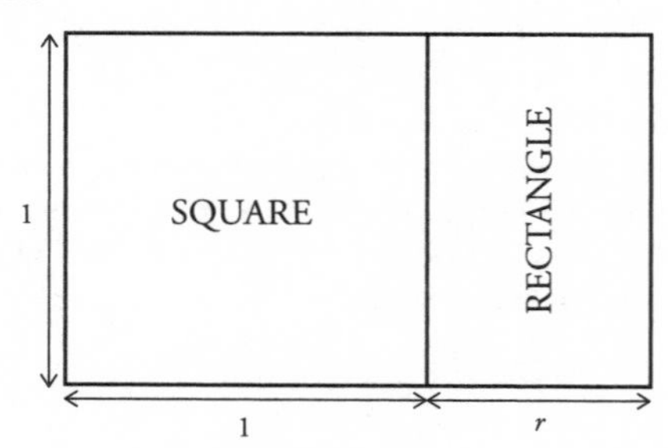

Consider a unit square, joined to a rectangle with sides of length r and 1. Then this rectangle's sides are in the ratio of r:1, and we want the same ratio to hold for the sides of the larger rectangle (i.e., the union of the square and rectangle). Since this larger rectangle has sides 1 and $1 + r$, we get the equation

$$\frac{1}{1+r} = \frac{r}{1}$$

which gives the quadratic equation $r^2 + r - 1 = 0$. The positive solution is $r = (-1 + \sqrt{5})/2$, from which the Golden Ratio $1 + r$ turns out to be $(1 + \sqrt{5})/2$.

204 ***phyllotaxis:*** For a mathematical model of this, see T. Okabe, "Biophysical Optimality of the Golden Angle in Phyllotaxis," *Scientific Reports* 5 (2015): 15358, https://doi.org/10.1038/srep15358.

206 ***Some scientific papers:*** The galaxy claim can be found in, e.g., R. Oldershaw, "The Preferred Pitch Angle of Spiral Galaxies: Mathematical and Physical Implications," *MNASSA* 41 (1982): 42–46.

207 ***falcons spiraling toward their prey:*** See Mario Livio, *The Golden Ratio: The Story of Phi, the World's Most Astonishing Number* (New York: Broadway, 2002).

230 ***fractional dimensionality:*** Here is how mathematicians calculate the fractional dimension d of an object that is self-similar. Imagine that you enlarge the object so that you retain the same shape, but each side is twice the original length. Then count the number of copies of the original contained in this enlarged version. If $d = 1$, i.e., the object is 1-D, then we know that this doubling will contain $2^1 = 2$ copies of the original (think of a line segment). For a 2-D object ($d = 2$), the enlarged version will contain $2^2 = 4$ copies of the object (think of a unit square, which when enlarged contains 4 identical squares). For a 3-D object ($d = 3$), you will have $2^3 = 8$ copies of the original (think of a unit cube, which when enlarged contains 8 identical cubes). So, in general, if an object has dimension d, then the enlarged version creates 2^d copies. This equation can be reversed to find the dimension, i.e., if we know that the enlarged object creates N copies, then d must be the solution of the equation $2^d = N$.

It can be easily seen that the Sierpinski triangle, when enlarged, contains 3 copies of its original self. In other words, $N = 3$. Solving $2^d = 3$ gives $d = \log 3 \div \log 2 \approx 1.585$. Hence the fractional dimension of the Sierpinski triangle is about 1.6.

234 ***a more sophisticated algorithm:*** Huseyin Kaya (2022). Fractal landscape generation with diamond-square algorithm (https://www.mathworks.com/matlabcentral/fileexchange/44714-fractal-landscape-generation-with-diamond-square-algorithm), MATLAB Central File Exchange.

237 ***fractal in nature:*** The fractal nature of rainfall distribution, both in terms of space and in terms of time, has been explored in S. Lovejoy and B. B. Mandelbrot, "Fractal Properties of Rain, and a Fractal Model," *Tellus A: Dynamic Meteorology and Oceanography* 37, no. 3 (1985): 209–32, https://doi.org/10.3402/tellusa.v37i3.11668.

239 ***the proposed mechanism:*** While there are alternative models for how shell pigmentation occurs, the one discussed here, taken from S. Wolfram, *A New Kind of Science* (Champaign, IL: Wolfram Research, 2002), and based on "cellular automata," is the simplest. Wolfram's book classifies and explores all 256 such possible rules; the one we discuss in detail here is rule 126.

240 ***a pattern with great complexity:*** the diagrams on this page were generated using a code by David Young (2022). Elementary Cellular Automata (https://www.mathworks.com/matlabcentral/fileexchange/26929-elementary-cellular-automata), MATLAB Central File Exchange.

240 ***different designs on shells:*** More complex models for shell pigmentation can generate even more patterns. These are generally based on reaction-diffusion differential equations, an idea first proposed by Alan Turing (see the second remark below under Day 7), and explored in detail in the landmark book by Hans Meinhardt, *The Algorithmic Beauty of Sea Shells*, 4th ed. (Berlin: Springer, 2009). Recently, models based on neural networks have also been proposed. See, e.g., A. Boettiger,

B. Ermentrout, and G. Oster, "The Neural Origins of Shell Structure and Pattern in Aquatic Mollusks," *Proceedings of the National Academy of Sciences USA* 106, no. 16 (2009), https://doi.org/10.1073/PNAS.0810311106.

Day 5: PHYSICS

255 ***practical role axioms played:*** The axiomatization of the mathematical branches of physics was the sixth of twenty-three open problems Hilbert announced in a famous address at the 1900 International Congress of Mathematicians in Paris. This problem was a natural extension of his work on axiomatizing geometry (see the endnote for page 96), and one he worked on for many years afterward.

The notion of axiomatizing their field was not universally accepted by physicists. Several argued that axioms were successful in math since the subject wasn't, strictly speaking, tied to anything. Its free-floating theorems might help explain the universe, but could also exist outside it. Physics, in contrast, was *entirely* about describing the universe. Its laws were beholden to reality, and had to be constantly fine-tuned to work in different ranges. Moreover, these laws had to evolve as more about the universe was discovered, so they could never enjoy the same absolute permanence as mathematical theorems. Experiments, not axioms, were what drove physics.

While these apprehensions may have had some validity, they did not directly contradict Hilbert's charge. His idea was to use axioms not for the initial construction of physics and its laws but to deduce them from what was known, and use them to shore up the subject's foundations where they were shaky. Such axioms could then resolve discrepancies, clarify interpretations, and put everything on firmer ground. They could help assess the compatibility of various laws, and also give insight for developing theories to come. Experiments would still be essential, since only through them could you compare different sets of axioms, and pick those that yielded laws conforming to physical reality. With these goals articulated, Hilbert was able to significantly influence the work of physicists like Max Born.

Since then, axiomatic theories have been developed for several parts of physics, and have been particularly successful in areas related to quantum mechanics. A number of these advances are described in the theme issue "Hilbert's Sixth Problem," compiled and edited by Luigi Accardi, Pierre Degond, and Alexander N. Gorban, *Philosophical Transactions of the Royal Society A* 376, no. 2118 (2018). As a side note, let me mention that Carnegie Mellon University, where I earned my PhD in 1983, had several faculty members (most notably, Walter Noll), whose research consisted of axiomatizing continuum mechanics.

257 ***alternatives:*** In addition to the many scientific and philosophical treatises on the subject, there is also a novel, *Einstein's Dreams* by Alan Lightman (New York: Pantheon, 1993), which looks at various interpretations of time through the lens of fiction.

263 ***Emmy Noether:*** The profundity of Noether's results becomes even more remarkable when considered against the gender discrimination she faced throughout most of her career. Her most well-known symmetry laws are usually phrased as follows: (1) symmetry under space translations gives rise to the conservation of linear momentum, (2) symmetry under time translations gives rise to the conservation of energy, (3) symmetry under rotations gives rise to the conservation of angular momentum. A summary of her work and life can be found in the book

review, Robyn Arianrhod, "The Evolution of an Idea," *Notices of the AMS* 60, no. 7 (August 2013): 916–19, http://dx.doi.org/10.1090/noti1027.

268 ***too complicated to include:*** One big simplification we made was related to spacetime, where we considered time as an extra dimension, similar to the three space dimensions, so that we had a normal "Euclidean" space (2-D in our case, but 4-D if space was three-dimensional). This doesn't work as a setting for relativity because objects cannot travel faster than *c*, the speed of light. So, for example, a straight trajectory through spacetime, like the ones we drew on page 259, would be realizable only if the speed involved (i.e., the reciprocal of the slope) was no more than *c*. Moreover, the concept of "distance" has to be modified in spacetime so that relativistic effects are accounted for. The most commonly used geometry to formulate relativity, developed by Minkowski, introduces different definitions of distance (and hence geodesics) between points, depending on where they lie in spacetime. It thus treats the time dimension very differently from space dimensions.

For more, see, for instance, the online notes for the course "Einstein for Everyone" by Professor John D. Norton of the University of Pittsburgh, https://www.pitt.edu/~jdnorton/teaching/HPS_0410/index.html.

270 ***a minuscule blip:*** The estimate comes from "If You Were to Move All of the Matter in the Universe into One Corner, How Much Space Would It Take Up?," HowStuffWorks.com, April 1, 2000, https://science.howstuffworks.com/dictionary/astronomy-terms/question221.htm.

Day 6: INFINITY

277 ***only rarely causes a problem:*** There have, however, been notable exceptions. See, for example, Kees Vuick, "Some Disasters Caused by Numerical Errors," http://ta.twi.tudelft.nl/users/vuik/wi211/disasters.html.

278 ***discrete space and time:*** See, for instance, Peter Forrest, "Is Space-Time Discrete or Continuous? An Empirical Question," *Synthese* 103, no. 3 (1995): 327–54, http://www.jstor.org/stable/20117405.

278 ***minimum length:*** See, for instance, Ethan Siegel, "What Is the Smallest Possible Distance in the Universe?," Forbes.com, June 26, 2019, https://www.forbes.com/sites/startswithabang/2019/06/26/what-is-the-smallest-possible-distance-in-the-universe/?sh=7c4d13c648a1.

295 ***mathematical work to Pope Leo XIII:*** See Joseph W. Dauben, "Georg Cantor and Pope Leo XIII: Mathematics, Theology, and the Infinite," *Journal of the History of Ideas* 38, no. 1 (January–March 1977): 85–108.

306 ***points in a line segment:*** The fact that the points in a line segment have a cardinality higher than the naturals leads me to my original *New York Times* article. At one point, I ask the reader to imagine what will happen if one draws a regular polygon with more and more sides: "Eventually, the sides shrink so much that the kinks start flattening out and the perimeter begins to appear curved. And then you see it: what will emerge is a circle, while at the same time the polygon can never actually become one." In actuality, you won't get a circle in the limit, since increasing the number of sides will lead you only to a polygon with $\aleph_0$ sides, while you would need *c* sides for a circle. The mathematical name for what you would get is a regular *apeirogon*. Of course, it's impossible to draw this figure in a way to distinguish it from a circle.

306 ***mostly filled with irrational numbers:*** As mentioned in the endnote for page 47 ("unrelated to any rational"), irrational numbers can be further divided into *algebraic* numbers like $\sqrt{2}$ (which can be expressed as the roots of polynomial equations with integer coefficients) and *transcendental* numbers like π (which cannot). One can show that the cardinality of algebraic numbers is again $\aleph_0$, which means the numbers responsible for the higher cardinality of the reals are the transcendentals. In other words, almost all numbers on the real line are transcendental, even though we would have a hard time even naming very many such specimens.

308 ***more involved argument:*** See, for instance, William Dunham, *Journey through Genius,* (New York: Penguin, 1991), 272–73.

309 ***Cantor's ingenious proof:*** Let S be a set with (any) cardinality card(S), and let $P(S)$ denote the set of all subsets of S (called the "power set" of S). Assume card($P(S)$) = card(S). Then we can establish a one-to-one correspondence between S and $P(S)$, i.e., for every element in S, there has to be one and exactly one element of $P(S)$, and vice versa. Hence, for each element a of S, there exists some subset of S that is uniquely matched to a. Call this subset match(a).

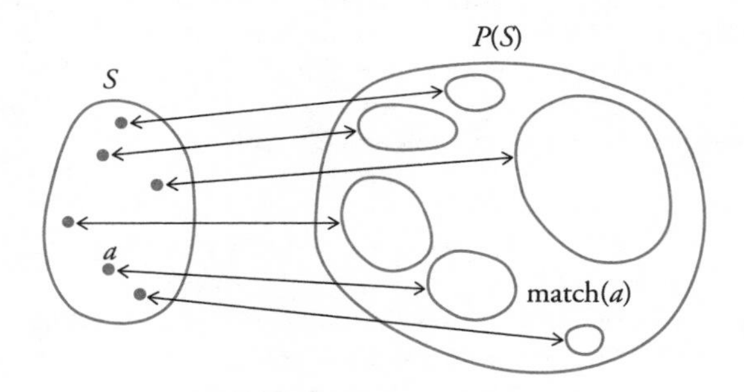

Now, since match(a) is a subset of S, and a is in S, either a belongs to match(a), or a does not belong to match(a). (For instance, if match(a) happens to be the singleton $\{a\}$, then clearly a belongs to match(a). On the other hand, if match(a) happens to be the empty set $\emptyset$, then obviously a does not belong to match(a).)

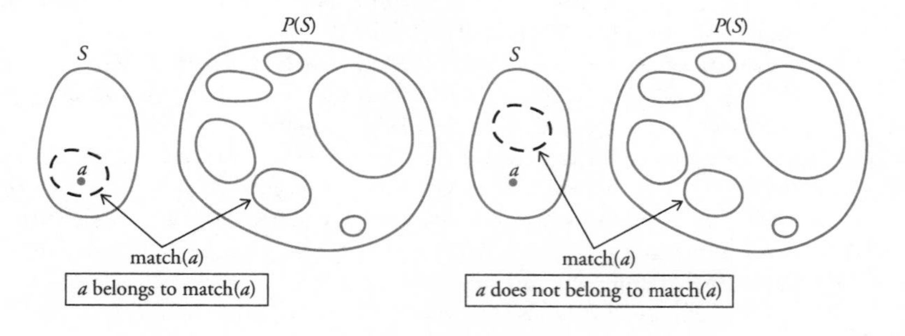

This means there are two types of elements in S. Call a an element of type 1 if it belongs to match(a), and of type 2 if it does not belong to match(a).

Now let B be the subset of all type 2 elements. Since B is an element of $P(S)$, there has to be an element b in S that is matched up with B, i.e., such that B equals match(b). This brings up the critical question: Does b belong to B or not?

Suppose yes. Then since B is the set of elements b of S that do not belong to match(b), b cannot belong to match(b) = B. So this gives a contradiction.

Suppose no, i.e., b does not belong to B. Then b does not belong to match(b). But this implies, by the definition of B, that b belongs to B, another contradiction.

The only possibility is that the assumption that card(S) = card($P(S)$) has to be incorrect. Since card($P(S)$) cannot be less than card(S), this implies card($P(S)$) > card(S).

311 ***giving up mathematics:*** Cantor spent much of his time after he gave up mathematics trying to prove that the plays by Shakespeare were written by Francis Bacon, a figure from the Renaissance best known for developing the scientific method.

311 ***resolve every question posable:*** As we've seen, mathematics has to be built up axiomatically. Depending on which axioms you take as your building blocks, you get different deductive (formal) systems of mathematics. In an earlier note, we called a set of axioms *consistent* if it does not lead to any logical contradictions. A deductive system is called *complete* if any statement that can be stated in the language of the system can be proved true or false using the axioms of the system. What Gödel proved was that a consistent deductive system that was advanced enough to support a certain amount of arithmetic could not be complete. In other words, you'll always have *undecidable* statements in such a system.

One of the most commonly used set of axioms for set theory, formulated by Zermelo and Fraenkel, is the ZFC system. Gödel proved, in 1940, that the continuum hypothesis could not be *disproved* by using these axioms. In 1963, another mathematician, Paul Cohen, established that the continuum hypothesis could not be *proved* using these axioms either. So the continuum hypothesis is an example of an undecidable statement in the ZFC system.

Intriguingly, just because a statement is undecidable doesn't mean it can't be either true or false. For instance, many mathematicians believe the continuum hypothesis is true (or false), but just can't be *proved* as such.

312 ***the root of everything:*** The mathematician David Hilbert believed in the ideal that mathematics could be shown to be consistent. This was the second of the twenty-three open mathematics problems he proposed in 1900 (his first was to prove or disprove the continuum hypothesis; his sixth, recall, was to axiomatize physics). Hilbert felt that all mathematical truths were within the reach of humans ("We must know. We will know," he declared at an address in Konigsberg, toward the end of his career). Gödel's incompleteness results showed that such consistency could not be established using the tools of the mathematical system itself. As such, Hilbert's ideal could not be realized in the expansive form in which it was intended. See, e.g., S. Budiansky, *Journey to the Edge of Reason: The Life of Kurt Gödel* (New York: W. W. Norton, 2021), chap. 5.

312 ***nonsense:*** The physics equivalent of such nonsense is to claim that the Heisenberg uncertainty principle, which says that we cannot exactly measure both a particle's position and velocity at the same instant, is the reason behind uncertainty in finance, art, philosophy, what-have-you.

Day 7: EMERGENCE

321 ***based on experimental evidence:*** A popular airing of this idea is given, for instance, in David Lindley, *The Dream Universe: How Fundamental Physics Lost Its Way* (New York: Doubleday, 2020).

325 ***spontaneous appearance:*** The picture of stripes forming is taken from Y. Suzuki, T. Takayama, I. N. Motoike, and T. Asai, "Striped and Spotted Pattern Generation on Reaction-Diffusion Cellular Automata—Theory and LSI Implementation," *International Journal of Unconventional Computing* (2007): 1713–19. The mathematical model tracks how pigmentation-enhancing *activators* interact with pigmentation-suppressing *inhibitors*. Initially, these two "morphogens" are assumed to be randomly mixed together, but as the animal grows, this distribution evolves under a two-dimensional "neighboring cells" rule (similar to the one-dimensional rule we explained earlier for the Sierpinski shell) to produce the pattern.

Such patterns are called Turing patterns in honor of Alan Turing, who first came up with a mathematical model for them in his landmark paper "The Chemical Basis of Morphogenesis," *Philosophical Transactions of the Royal Society of London. Series B, Biological Sciences* 237, no. 641 (1952): 37–72. Turing's work has been used as the basis of modeling pattern formation in many different contexts, including fingerprints. See, e.g., D. A. Garzón-Alvarado and A. M. Ramírez Martinez, "A Biochemical Hypothesis on the Formation of Fingerprints Using a Turing Patterns Approach," *Theoretical Biology and Medical Modelling* 8 (2011): article 24, https://doi.org/10.1186/1742-4682-8-24.

325 ***complexity theory:*** Two classic references I'd recommend for complexity theory and emergence are G. Flake, *The Computational Beauty of Nature* (Cambridge: MIT Press, 1998) and M. Waldrop, *Complexity: The Emerging Science at the Edge of Order and Chaos* (New York: Simon & Schuster, 1992). The first gives an introduction to various computer simulations that help explain what emergence is about, while the second gives an account of the early history of complexity and emergence. The second book's subtitle refers to the observation that the most interesting emergent behavior is usually found in a narrow range where the parameters of a system make it neither orderly (in which case its evolution would be stable and predictable) nor chaotic (in which case disorder would obscure everything).

327 ***ant bridges:*** These are described in J. Graham, A. Kao, D. Wilhelm, and S. Garnier, "Optimal Construction of Army Ant Living Bridges," *Journal of Theoretical Biology* 435 (2017): 184–98, https://doi.org/10.1016/j.jtbi.2017.09.017. A good reference for general insect intelligence (which describes the mechanism behind optimizing of ant paths) is B. Hölldobler and E. O. Wilson, *The Superorganism: The Beauty, Elegance, and Strangeness of Insect Societies* (New York: W. W. Norton, 2009).

ILLUSTRATION CREDITS

Diagrams and drawings:

The primary illustrator for technical diagrams is Dae In Chung. Additional artwork is by José Villarrubia. All diagrams and drawings are based on initial renditions by the author; artwork on p. 36 is by author.

Photo and image credits:

71: Vassily Kandinsky (Russian, 1866–1944). *Kleine Welten IV (Small Worlds IV)*, 1922. Color lithograph, image area: $11\frac{7}{16} \times 10\frac{3}{16}$ inches (29.05 × 25.88 cm); sheet: $14\frac{1}{4} \times 11\frac{3}{8}$ inches (36.19 × 28.83 cm). Collection Albright-Knox Art Gallery, Buffalo, New York; Gift of Frederic P. Norton, 1999 (P1999:6.280). Photo: Tom Loonan for Albright-Knox Art Gallery.

113–115 (top), 116 (top): Images courtesy of Shelley Husband.

117, 119 (bottom): Images courtesy of Daina Taimina.

148: *The Scream* by Edvard Munch, 1893. Wikimedia Commons / Public Domain.

174 (top): Rabbit mesh image courtesy of Dr. Sebastian Valette, CREATIS, CNRS, France. Image created using ACVD software, see S. Valette, J. M. Chassery, R. Prost, "Generic remeshing of 3D triangular meshes with metric-dependent discrete Voronoi Diagrams," *IEEE Transactions on Visualization and Computer Graphics*, 14 , No. 2 (2008): 369–81.

211–214: *The Mona Lisa* by Leonardo da Vinci, between 1503 and 1506. Wikimedia Commons / Public Domain.

226 (bottom left), 229 (center), 230: Images courtesy of Dr. Ted Burke, Technological University Dublin.

226 (bottom right), 232, 236: Images by Wernher Krutein/Photovault.com.

233: Illustration created by Larry Cole, based on map services and data available from U.S. Geological Survey, National Geospatial Program.

237: Image courtesy of Larry Cole.

268: Image courtesy of NASA & ESA.

325: Image from Y. Suzuki, T. Takayama, I. N. Motoike and T. Asai., “Striped and Spotted Pattern Generation on Reaction-diffusion Cellular Automata–Theory and LSI Implementation,” *IJUC* Vol 3, no. 1 (2007): 1-13. Courtesy of Old City Publishing.

INDEX

Page numbers in *italics* indicate a figure, and page numbers in **bold** indicate a table on the corresponding page.